INTELLIGENCE AND COUNTERINTELLIGENCE STUDIES

SECURITY ISSUES AND CHINESE TELECOMMUNICATIONS COMPANIES

HUAWEI AND ZTE

INTELLIGENCE AND COUNTERINTELLIGENCE STUDIES

Additional books in this series can be found on Nova's website under the Series tab.

Additional E-books in this series can be found on Nova's website under the E-book tab.

DEFENSE, SECURITY AND STRATEGIES

Additional books in this series can be found on Nova's website under the Series tab.

Additional E-books in this series can be found on Nova's website under the E-book tab.

INTELLIGENCE AND COUNTERINTELLIGENCE STUDIES

SECURITY ISSUES AND CHINESE TELECOMMUNICATIONS COMPANIES

HUAWEI AND ZTE

SIMON JOHNSTON
EDITOR

nova publishers
New York

For permission to use material from this book please contact us:
Telephone 631-231-7269; Fax 631-231-8175
Web Site: http://www.novapublishers.com

Additional color graphics may be available in the e-book version of this book.

Library of Congress Cataloging-in-Publication Data

ISBN: 978-1-62417-423-0

Published by Nova Science Publishers, Inc. † New York

CONTENTS

PREFACE

The House Permanent Select Committee on Intelligence initiated an investigation in November 2011 to inquire into the counterintelligence and security threat posed by Chinese telecommunications companies doing business in the United States. Prior to initiating the formal investigation, the Committee performed a preliminary review of the issue, which confirmed significant gaps in available information about the Chinese telecommunications sector, the histories and operations of specific companies operating the U.S., and those companies' potential ties to the Chinese state. This book examines the potential security threat posed by Chinese telecommunications companies with potential ties to the Chinese government or military. In particular, to the extent these companies are influenced by the state, or provide Chinese intelligence services access to telecommunication networks, the opportunity exists for further economic and foreign espionage by a foreign nation-state already known to be a major perpetrator of cyber espionage.

Chapter 1 – In February 2011, Huawei Technologies Company, the leading Chinese telecommunications equipment manufacturer, published an open letter to the U.S. Government denying security concerns with the company or its equipment, and requesting a full investigation into its corporate operations. Huawei apparently believed – correctly – that without a full investigation into its corporate activities, the United States could not trust its equipment and services in U.S. telecommunications networks.

The House Permanent Select Committee on Intelligence (herein referred to as "the Committee" initiated this investigation in November 2011 to inquire into the counterintelligence and security threat posed by Chinese telecommunications companies doing business in the United States. Prior to initiating the formal investigation, the Committee performed a preliminary review of the issue, which confirmed significant gaps in available information about the Chinese telecommunications sector, the histories and operations of specific companies operating in the United States, and those companies' potential ties to the Chinese state. Most importantly, that preliminary review highlighted the potential security threat posed by Chinese telecommunications companies with potential ties to the Chinese government or military. In particular, to the extent these companies are influenced by the state, or provide Chinese intelligence services access to telecommunication networks, the opportunity exists for further economic and foreign espionage by a foreign nation-state already known to be a major perpetrator of cyber espionage.

As many other countries show through their actions, the Committee believes the telecommunications sector plays a critical role in the safety and security of our nation, and is thus a target of foreign intelligence services. The Committee's formal investigation focused on Huawei and ZTE, the top two Chinese telecommunications equipment manufacturers, as they seek to market their equipment to U.S. telecommunications infrastructure. The Committee's main goal was to better understand the level of risk posed to the United States as these companies hope to expand in the United States. To evaluate the threat, the investigation involved two distinct yet connected parts: (1) a review of open-source information on the Companies' histories, operations, financial information, and potential ties to the Chinese government or Chinese Communist Party; and (2) a review of classified information, including a review of programs and efforts of the U.S. Intelligence Community (IC) to ascertain whether the IC is appropriately prioritizing and resourced for supply chain risk evaluation.

Despite hours of interviews, extensive and repeated document requests, a review of open-source information, and an open hearing with witnesses from both companies, the Committee remains unsatisfied with the level of cooperation and candor provided by each company. Neither company was willing to provide sufficient evidence to ameliorate the Committee's concerns. Neither company was forthcoming with detailed information about its formal relationships or regulatory interaction with Chinese authorities. Neither company provided specific details about the precise role of each company's Chinese Communist Party Committee. Furthermore, neither company provided detailed information about its operations in the United States. Huawei, in particular, failed to provide thorough information about its corporate structure, history, ownership, operations, financial arrangements, or management. Most importantly, neither company provided sufficient internal documentation or other evidence to support the limited answers they did provide to Committee investigators.

During the investigation, the Committee received information from industry experts and current and former Huawei employees suggesting that Huawei, in particular, may be violating United States laws. These allegations describe a company that has not followed United States legal obligations or international standards of business behavior. The Committee will be referring these allegations to Executive Branch agencies for further review, including possible investigation.

In sum, the Committee finds that the companies failed to provide evidence that would satisfy any fair and full investigation. Although this alone does not prove wrongdoing, it factors into the Committee's conclusion below. Further, this report contains a classified annex, which also adds to the Committee's concerns about the risk to the United States. The investigation concludes that the risks associated with Huawei's and ZTE's provision of equipment to U.S. critical infrastructure could undermine core U.S. national-security interests.

Chapter 2 – This is the Statement of Chairman Mike Rogers, House Permanent Select Committee on Intelligence. Hearing on "Investigation of the Security Threat Posed by Chinese Telecommunications Companies Huawei and ZTE."

Chapter 3 – This is the Statement of Dutch Ruppersberger, Ranking Member, House Permanent Select Committee on Intelligence. Hearing on "Investigation of the Security Threat Posed by Chinese Telecommunications Companies Huawei and ZTE."

Chapter 4 – This is the Statement of Charles Ding, Corporate Senior Vice President, Huawei. Hearing on "Investigation of the Security Threat Posed by Chinese Telecommunications Companies Huawei and ZTE."

Chapter 5 – This is the Statement of Zhu Jinyun, Senior Vice President for Europe and North America, ZTE. Hearing on "Investigation of the Security Threat Posed by Chinese Telecommunications Companies Huawei and ZTE."

Chapter 6 – Foreign economic collection and industrial espionage against the United States represent significant and growing threats to the nation's prosperity and security.

Cyberspace—where most business activity and development of new ideas now takes place—amplifies these threats by making it possible for malicious actors, whether they are corrupted insiders or foreign intelligence services (FIS), to quickly steal and transfer massive quantities of data while remaining anonymous and hard to detect.

In: Security Issues …
Editor: Simon Johnston

ISBN: 978-1-62417-423-0

Chapter 1

INVESTIGATIVE REPORT ON THE U.S. NATIONAL SECURITY ISSUES POSED BY CHINESE TELECOMMUNICATIONS COMPANIES HUAWEI AND ZTE*

U.S. House of Representatives

EXECUTIVE SUMMARY

In February 2011, Huawei Technologies Company, the leading Chinese telecommunications equipment manufacturer, published an open letter to the U.S. Government denying security concerns with the company or its equipment, and requesting a full investigation into its corporate operations.[1] Huawei apparently believed – correctly – that without a full investigation into its corporate activities, the United States could not trust its equipment and services in U.S. telecommunications networks.[2]

The House Permanent Select Committee on Intelligence (herein referred to as "the Committee" initiated this investigation in November 2011 to inquire into the counterintelligence and security threat posed by Chinese telecommunications companies doing business in the United States. Prior to initiating the formal investigation, the Committee performed a preliminary

* This is an edited, reformatted and augmented version of the House Permanent Select Committee on Intelligence, dated October 8, 2012.

review of the issue, which confirmed significant gaps in available information about the Chinese telecommunications sector, the histories and operations of specific companies operating in the United States, and those companies' potential ties to the Chinese state. Most importantly, that preliminary review highlighted the potential security threat posed by Chinese telecommunications companies with potential ties to the Chinese government or military. In particular, to the extent these companies are influenced by the state, or provide Chinese intelligence services access to telecommunication networks, the opportunity exists for further economic and foreign espionage by a foreign nation-state already known to be a major perpetrator of cyber espionage.

As many other countries show through their actions, the Committee believes the telecommunications sector plays a critical role in the safety and security of our nation, and is thus a target of foreign intelligence services. The Committee's formal investigation focused on Huawei and ZTE, the top two Chinese telecommunications equipment manufacturers, as they seek to market their equipment to U.S. telecommunications infrastructure. The Committee's main goal was to better understand the level of risk posed to the United States as these companies hope to expand in the United States. To evaluate the threat, the investigation involved two distinct yet connected parts: (1) a review of open-source information on the Companies' histories, operations, financial information, and potential ties to the Chinese government or Chinese Communist Party; and (2) a review of classified information, including a review of programs and efforts of the U.S. Intelligence Community (IC) to ascertain whether the IC is appropriately prioritizing and resourced for supply chain risk evaluation.[3]

Despite hours of interviews, extensive and repeated document requests, a review of open-source information, and an open hearing with witnesses from both companies, the Committee remains unsatisfied with the level of cooperation and candor provided by each company. Neither company was willing to provide sufficient evidence to ameliorate the Committee's concerns. Neither company was forthcoming with detailed information about its formal relationships or regulatory interaction with Chinese authorities. Neither company provided specific details about the precise role of each company's Chinese Communist Party Committee. Furthermore, neither company provided detailed information about its operations in the United States. Huawei, in particular, failed to provide thorough information about its corporate structure, history, ownership, operations, financial arrangements, or management. Most importantly, neither company provided sufficient internal documentation or other evidence to support the limited answers they did provide to Committee investigators.

During the investigation, the Committee received information from industry experts and current and former Huawei employees suggesting that Huawei, in particular, may be violating United States laws. These allegations describe a company that has not followed United States legal obligations or international standards of business behavior. The Committee will be referring these allegations to Executive Branch agencies for further review, including possible investigation.

In sum, the Committee finds that the companies failed to provide evidence that would satisfy any fair and full investigation. Although this alone does not prove wrongdoing, it factors into the Committee's conclusion below. Further, this report contains a classified annex, which also adds to the Committee's concerns about the risk to the United States. The investigation concludes that the risks associated with Huawei's and ZTE's provision of equipment to U.S. critical infrastructure could undermine core U.S. national-security interests.

Based on this investigation, the Committee provides the following recommendations:

Recommendation 1: The United States should view with suspicion the continued penetration of the U.S. telecommunications market by Chinese telecommunications companies.

- The United States Intelligence Community (IC) must remain vigilant and focused on this threat. The IC should actively seek to keep cleared private sector actors as informed of the threat as possible.
- The Committee on Foreign Investment in the United States (CFIUS) must block acquisitions, takeovers, or mergers involving Huawei and ZTE given the threat to U.S. national security interests. Legislative proposals seeking to expand CFIUS to include purchasing agreements should receive thorough consideration by relevant Congressional committees.
- U.S. government systems, particularly sensitive systems, should not include Huawei or ZTE equipment, including component parts. Similarly, government contractors ± particularly those working on contracts for sensitive U.S. programs ± should exclude ZTE or Huawei equipment in their systems.

Recommendation 2: Private-sector entities in the United States are strongly encouraged to consider the long-term security risks associated with doing business with either ZTE or Huawei for equipment or services. U.S. network providers and systems developers are strongly encouraged to seek other vendors for their projects. Based on available classified and unclassified information, Huawei and ZTE cannot be trusted to be free of foreign state influence and thus pose a security threat to the United States and to our systems.

Recommendation 3: Committees of jurisdiction within the U.S. Congress and enforcement agencies within the Executive Branch should investigate the unfair trade practices of the Chinese telecommunications sector, paying particular attention to China's continued financial support for key companies.

Recommendation 4: Chinese companies should quickly become more open and transparent, including listing on a western stock exchange with advanced transparency requirements, offering more consistent review by independent third-party evaluators of their financial information and cyber-security processes, complying with U.S. legal standards of information and evidentiary production, and obeying all intellectual-property laws and standards. Huawei, in particular, must become more transparent and responsive to U.S. legal obligations.

Recommendation 5: Committees of jurisdiction in the U.S. Congress should consider potential legislation to better address the risk posed by telecommunications companies with nation-state ties or otherwise not clearly trusted to build critical infrastructure. Such legislation could include increasing information sharing among private sector entities, and an expanded role for the CFIUS process to include purchasing agreements.

Report

I. The threat posed to U.S. national-security interests by vulnerabilities in the telecommunications supply chain is an increasing priority given: the country's reliance on interdependent critical infrastructure systems; the range of threats these systems face; the rise in cyber espionage; and the growing dependence all consumers have on a small group of equipment providers.

The United States' critical infrastructure, and in particular its telecommunications networks, depend on trust and reliability. Telecommunications networks are vulnerable to malicious and evolving intrusions or disruptive activities. A sufficient level of trust, therefore, with both the provider of the equipment and those performing managed services must exist at all times. A company providing such equipment, and particularly any company having access to or detailed knowledge of the infrastructures' architectural blueprints, must be trusted to comply with United States laws, policies, and standards. If it cannot be trusted, then the United States and others should question whether the company should operate within the networks of our critical infrastructure.

The risk posed to U.S. national-security and economic interests by cyber-threats is an undeniable priority. First, the country's reliance on telecommunications infrastructure includes more than consumers' use of computer systems. Rather, multiple critical infrastructure systems depend on information transmission through telecommunications systems. These modern critical infrastructures include electric power grids; banking and finance systems; natural gas, oil, and water systems; and rail and shipping channels; each of which depend on computerized control systems. Further, system interdependencies among these critical infrastructures greatly increase the risk that failure in one system will cause failures or disruptions in multiple critical infrastructure systems.[4] Therefore, a disruption in telecommunication networks can have devastating effects on all aspects of modern American living, causing shortages and stoppages that ripple throughout society.

Second, the security vulnerabilities that come along with this dependence are quite broad, and range from insider threats[5] to cyber espionage and attacks from sophisticated nation states. In fact, there is a growing recognition of vulnerabilities resulting from foreign-sourced telecommunications supply chains used for U.S. national-security applications. The FBI, for example, has assessed with high confidence that threats to the supply chain from both nation-states and criminal elements constitute a high cyber threat.[6] Similarly, the National Counterintelligence Executive assessed that "foreign attempts to collect U.S. technological and economic information will continue at a high level and will represent a growing and persistent threat to US economic security."[7]

Third, the U.S. government must pay particular attention to products produced by companies with ties to regimes that present the highest and most advanced espionage threats to the U.S., such as China. Recent cyber-attacks often emanate from China, and even though precise attribution is a perennial challenge, the volume, scale, and sophistication often indicate state involvement. As the U.S.-China Commission explained in its unclassified report on China's capabilities to conduct cyber warfare and computer network exploitation (CNE), actors in China seeking sensitive economic and national security information through malicious cyber operations often face little chance of being detected by their targets.[8]

Finally, complicating this problem is the fact that Chinese telecommunications firms, such as Huawei and ZTE, are rapidly becoming dominant global players in the telecommunications market. In another industry, this development might not be particularly concerning. When those companies seek to control the market for sensitive equipment and infrastructure that could be used for spying and other malicious purposes, the lack of market diversity becomes a national concern for the United States and other countries.[9] Of note, the United States is not the only country focusing on these concerns. Australia expressed similar concerns when it chose to ban Huawei from its national broadband infrastructure project.[10] Great Britain has attempted to address the concerns by instituting an evaluation regime that limits Huawei's access to the infrastructure and evaluates any Huawei equipment and software before they enter the infrastructure.[11]

A. China has the means, opportunity, and motive to use telecommunications companies for malicious purposes.

Chinese intelligence collection efforts against the U.S. government are growing in "scale, intensity and sophistication."[12] Chinese actors are also the world's most active and persistent perpetrators of economic espionage.[13] U.S. private sector firms and cyber-security specialists report an ongoing onslaught of sophisticated computer network intrusions that originate in China, and are almost certainly the work of, or have the backing of, the Chinese government.[14] Further, Chinese intelligence services, as well as private companies and other entities, often recruit those with direct access to corporate networks to steal trade secrets and other sensitive proprietary data.[15]

These cyber and human-enabled espionage efforts often exhibit sophisticated technological capabilities, and these capabilities have the potential to translate into efforts to insert malicious hardware or software implants into Chinese-manufactured telecommunications components and systems marketed to the United States. Opportunities to tamper with telecommunications components and systems are present throughout product development, and vertically integrated industry giants like Huawei and ZTE provide a wealth of opportunities for Chinese intelligence agencies to insert malicious hardware or software implants into critical telecommunications components and systems.[16] China may seek cooperation from the leadership of a company like Huawei or ZTE for these reasons Even if the company's leadership refused such a request, Chinese intelligence services need only recruit working-level technicians or managers in these companies. Further, it appears that under Chinese law, ZTE and Huawei would be obligated to cooperate with any request by the Chinese government to use their systems or access them for malicious purposes under the guise of state security.[17]

A sophisticated nation-state actor like China has the motivation to tamper with the global telecommunications supply chain, with the United States being a significant priority. The ability to deny service or disrupt global systems allows a foreign entity the opportunity to exert pressure or control over critical infrastructure on which the country is dependent. The capacity to maliciously modify or steal information from government and corporate entities provides China access to expensive and time-consuming research and development that advances China's economic place in the world. Access to U.S. telecommunications infrastructure also allows China to engage in undetected espionage against the United States government and private sector interests.[18] China's military and intelligence services, recognizing the technological superiority of the U.S. military, are actively searching for asymmetrical advantages that could be exploited in any future conflict with the United States.[19] Inserting malicious hardware or software implants into Chinese-manufactured telecommunications components and systems headed for U.S. customers could allow Beijing to shut down or degrade critical national security systems in a time of crisis or war. Malicious implants in the components of critical infrastructure, such as power grids or financial networks, would also be a tremendous weapon in China's arsenal.

Malicious Chinese hardware or software implants would also be a potent espionage tool for penetrating sensitive U.S. national security systems, as well as providing access to the closed American corporate networks that contain the sensitive trade secrets, advanced research and development data, and negotiating or litigation positions that China would find useful in obtaining an unfair diplomatic or commercial advantage over the United States.

In addition to supply chain risks associated with hardware and software, managed services also pose a threat. Managed services, sold as part of the systems maintenance contract, allow for remote network access for everyday updates to software and patches to glitches. Unfortunately, such contracts may also allow the managed-service contractor to use its authorized access for malicious activity under the guise of legitimate assistance. Such access also offers an opportunity for more-tailored economic or state-sponsored espionage activities. Telecommunications companies such as Huawei are seeking to expand service portions of their business.[20]

The U.S. Government has acknowledged these concerns with telecommunications supply chain risk for several years. Indeed, as one of twelve critical infrastructure protection priorities outlined in the White House's 2009 Comprehensive National Cybersecurity Initiative (CNCI), Supply Chain Risk Management (SCRM) is identified as a national concern. Similarly, the Executive Branch continues to review supply chain issues consistent with its National Strategy for Global Supply Chain Security, released in January 2012. A key part of the management of supply chain risk, as explained in the report, is to properly "understand and identify vulnerabilities to the supply chain that stem from both exploitation of the system by those seeking to introduce harmful products or materials and disruptions from intentional attacks, accidents, or natural disasters."[21]

B. Suggested "mitigation measures" cannot fully address the threat posed by Chinese telecommunications companies providing equipment and services to United States critical infrastructure.

Many countries struggle with the potential threats posed by untrustworthy telecommunications companies. In Great Britain, the government took initial steps (as part of an overall mitigation strategy) to address its concerns by entering into an agreement with Huawei to establish an independently managed Cyber Security Evaluation Centre (CSEC). CSEC conducts independent reviews of Huawei's equipment and software deployed to the United Kingdom's telecommunications infrastructure, and provides such results to the relevant UK carriers and UK government. The goal of the British government is to attempt to lessen the threat that Huawei products deployed in critical UK telecommunications infrastructure pose to the availability or integrity of UK networks.

Huawei and ZTE have proposed similar schemes for products entering the United States market, administered without U.S. government involvement, but by Electronic Warfare Associates or other private-sector security firms.[22] These partnerships seek to address concerns that the companies could permit the Chinese government to insert features or vulnerabilities into their products, which would assist espionage or cyber warfare. Unfortunately, there are concerns that such efforts if taken in the United States will fall short of addressing security concerns given the breadth and scale of the U.S. telecommunications market.

Post-production evaluation processes are a standard approach to determining the security properties of complex, software-intensive systems. Processes like the Common Criteria for Information Technology Security Evaluation and various private certification services define a process by which an evaluator measures a product against a set of standards and assigns a security rating. The rating is meant to help a consumer know how much confidence to place in the security features of the device or software package. Both the implementation of the system and the methodology used to develop it, as documented by the manufacturer, are typically used as evidence for the chosen rating. Further, such processes are not necessarily designed to uncover malicious code but to encourage a foundational security baseline in security-enabled products.

For a variety of technical and economic reasons, evaluation programs as proposed by Huawei and ZTE are less useful than one might expect. In fact, the programs may create a false sense of security that an incomplete, flawed, or misapplied evaluation would provide. An otherwise careful consumer may choose to forego a thorough threat, application, and environment-based risk assessment, and the costs such evaluations entail, because an accredited outside expert has "blessed" the product in some way.

One key issue not addressed by standardized third-party security evaluations is product and deployment diversity. The behavior of a device or system can vary wildly depending on how and where it is configured, installed, and maintained. For time and cost reasons, an evaluation usually targets a snapshot of one product model configured in a specific and often unrealistically restrictive way. The pace of technology development today drives products to evolve far more rapidly than any third-party comprehensive evaluation process can follow. The narrow configuration specification used during testing almost ensures that an evaluated device will be deployed in a way not specifically covered by a formal evaluation. A security evaluation of a complex device is useless if the device is not deployed precisely in the same configuration as it was tested.

The evaluation of products prior to deployment only addresses the product portion of the lifecycle of networks. It is also important to recognize that how a network operator oversees its patch management, its trouble-shooting and maintenance, upgrades, and managed-service elements, as well as the vendors it chooses for such services, will affect the ongoing security of the network.

Vendors financing their own security evaluations create conflicts of interest that lead to skepticism about the independence and rigor of the result. A product manufacturer will naturally pursue its own interests and ends which are not necessarily aligned with all interests of the consumers. A different, but related, race to the bottom has been noted for the similarly vendor-financed Common Criteria evaluations.[23] The designers of the Common Criteria system understood this danger and implemented government certification for evaluators. The precaution seems mostly cosmetic as no certification has ever been challenged or revoked despite gaming and poor evaluation performance. Given similar concerns and about conflicts of interest, Huawei's U.K. evaluators of Huawei equipment have been vetted by the U.K. government and hold government security clearances, and the U.K. process has the support of the U.K. Carriers. It is not clear yet, however, that such steps would readily transfer to the U.S. market or successfully overcome the natural incentives of the situation and lead to truly independent investigations.

The task of finding and eliminating every significant vulnerability from a complex product is monumental. If we also consider flaws intentionally inserted by a determined and clever insider, the task becomes virtually impossible.[24] While there is a large body of literature describing techniques for finding latent vulnerabilities in hardware and software systems, no such technique claims the ability to find all such vulnerabilities in a pre-existing system. Techniques do exist that can prove a system implementation matches a design which has been formally verified to be free of certain types of flaws.[25] However, such formal techniques must be incorporated throughout the design and development process to be effective. They cannot currently be applied to a finished product of significant size or complexity. Even when embedded into a design and development process, formal techniques of this type do not yet scale to the size of complete commercial telecommunication systems. It is highly unlikely that a security evaluation partnership such as that proposed by Huawei or ZTE, independent of its competence and motives, will be able to identify all relevant flaws in products the size and complexity of core network infrastructure devices. If significant flaws remain in widely fielded products and processes that are known to a potential adversary, it seems like the evaluation process has provided only marginal benefit.

A security evaluation of potentially suspect equipment being deployed in critical infrastructure roles may seem like an answer to the security problems posed. Unfortunately, given the complexity of the telecommunications grid, the limitations of current security evaluation techniques, and the economics of vendor-financed analyses provide a sense of security but not actual security.

Significant security is available only through a thoughtful design and engineering process that addresses a complete system-of-systems across its full lifecycle, from design to retirement and includes aspects such as discrete technology components, their interactions, the human environment, and threats from the full spectrum of adversaries. The result of such a process should be a convincing set of diverse evidence that a system is worthy of our trust.[26]

II. Investigation

A. Scope of Investigation

The House Permanent Select Committee on Intelligence is responsible for authorizing the intelligence activities of the United States and overseeing those activities to ensure that they are legal, effective, and properly resourced to protect the national security interests of the United States. Specifically, the Committee is charged with reviewing and studying on a continuing basis the authorities, programs, and activities of the Intelligence Community and with reviewing and studying on an exclusive basis the sources and methods of the community.[27] Along with this responsibility is the obligation to study and understand the range of foreign threats faced by the United States, including those directed against our nation's critical infrastructure. Similarly, the Committee must evaluate the threats from foreign intelligence operations and ensure that the country's counterintelligence agencies are appropriately focused on and resourced to defeat those operations.[28]

The Committee's goals in this investigation were to inquire into the potential security risk posed by the top two Chinese telecommunications companies and review whether our government is properly positioned to understand and respond to that threat. An additional aim of this process has been to determine what information could be provided in an unclassified form to shed light on the key questions of whether the existence of these firms in our market would pose a national-security risk through the potential loss of control of U.S. critical infrastructure.

Of course, the United States' telecommunications sector increasingly relies on a global supply chain for the production and delivery of equipment and services. That reliance presents significant risks that other individuals or entities – including those backed by foreign governments – can and will exploit and undermine the reliability of the networks. Better understanding the supply-chain risks we face is vital if we are to protect the security and functionality of our networks and if we are to guard against national security and economic threats to those networks. The investigation's scope reflects the underlying need for the U.S. to manage the global supply chain system using a risk-based approach.

Recent studies highlight that actors in China are the source of more cyber-attacks than in any other country. The National Counterintelligence Executive, for example, explained, in an open report on cyber-espionage, "Chinese actors are the world's most active and persistent perpetrators of economic espionage."[29] Thus, the Committee focused on those companies with the strongest potential Chinese ties and those that also seek greater entry into the United States market. Both Huawei and ZTE are indigenous Chinese firms, with histories that include connections to the Chinese government. Both Huawei and ZTE have already incorporated United States' subsidiaries, and both are seeking to expand their footprint in the United States market. Huawei has received, thus far, the greatest attention from analysts and the media. Given the similarities of the two companies, however, including their potential ties to the Chinese government, support by the Chinese government, and goals to further their U.S. presence, the Committee focused on both Huawei and ZTE.

Both Huawei and ZTE assert that the Committee should not focus only on them to the exclusion of all other companies that manufacture parts or equipment in China. The Committee recognizes that many non-Chinese companies, including U.S. technology companies, manufacture some of their products in China. But it is not only the location of manufacturing that is important to the risk calculation. It is also ownership, history, and the products being marketed. These may not be the only two companies presenting this risk, but they are the two largest Chinese-founded, Chinese-owned telecommunications companies seeking to market critical network equipment to the United States. To review supply chain risk, the Committee decided to focus first on the largest perceived vulnerabilities, with an expectation that the conclusion of this investigation would inform how to view the potential threat to the supply chain from other companies or manufacturers operating in China and other countries.

B. Investigative Process

The Committee's investigative process included extensive interviews with company and government officials, numerous document requests, and an open hearing with a senior official from both Huawei and ZTE. Committee staff reviewed available information on the specific companies, and Committee staff and members met with Huawei and ZTE officials for lengthy, in-depth meetings and interviews. Committee staff also toured the companies' facilities and factories.

Specifically, on February 23, 2012, Committee staff met with and interviewed corporate executives of Huawei at its corporate headquarters in Shenzhen, China. The delegation toured Huawei's corporate headquarters, reviewed company product lines, and toured a large manufacturing factory. Officials involved in the discussion from Huawei included Ken Hu, Huawei's Deputy Chairman of the Board and Acting CEO; Evan Bai, Vice President of the Treasury Management Office; Charlie Chen, Senior Vice President in charge of Huawei (USA); Jiang Xisheng, Secretary of the Board; John Suffolk, Global Security Officer; and Rose Hao, Export Regulator.

On April 12, 2012, Committee staff met with and interviewed corporate executives of ZTE at its corporate headquarters in Shenzhen, China. In addition to these meetings, the delegation took a brief tour of ZTE's corporate headquarters, including a factory site. Officials from ZTE included Zhu Jinyun, ZTE's Senior Vice President, U.S. and North America Market; Fan Qingfeng, Executive Vice President of Global Marketing and Sales; Guo Jianjun, Legal Director; Timothy Steinert, Independent Director of the Board (and Ali Baba Counsel); Ma Xuexing, Legal Director; Cao Wei, Security and Investor Relations with the Information Disclosure Office; Qian Yu, Security and Investor Relations with the Information Disclosure Office; and John Merrigan, attorney with DLA Piper.

In May of 2012, Ranking Member Ruppersberger along with Committee members Representative Nunes, Representative Bachmann, and Representative Schiff traveled to Hong Kong to meet with senior officials of both Huawei and ZTE. In addition to the senior officials present at the staff meetings, the Committee members met with Ren Zhengfei, the founder and President of Huawei.

After the meetings, the Committee followed-up with several pages of written questions and document requests to fill in information gaps, inconsistent or incomplete answers, and the need for corroborating documentary evidence of many of the companies' factual and historical assertions. Unfortunately, neither company was completely or fully responsive to the Committee's document requests. Indeed, neither Huawei nor ZTE provided internal documents in response to the Committee's letter.[30] To attempt, again, to answer the remaining questions, the Committee called each company to an open hearing.

On September 13, 2012, the Committee held an open hearing with representatives of both ZTE and Huawei. The witnesses included Mr. Charles Ding, corporate senior vice president and Huawei's representative to the United States, and Mr. Zhu Jinyun, ZTE senior vice president for North America and Europe. The hearing was designed to be both thorough and fair. Thc witnesses were each given twenty minutes for an opening statement and each was aided by an interpreter during the question and answer portion of the hearing to ensure that the witnesses were given the maximum opportunity to understand the questions and answer completely and factually.[31]

Once again, the witnesses' answers were often vague and incomplete. For example, they claimed to have no understanding or knowledge of commonly used terms, could not answer questions about the composition of their internal Party Committees, refused to provide straightforward answers about their operations in the United States, sought to avoid answering questions about their histories of intellectual property protection, and claimed to have no understanding or knowledge of Chinese laws that force them to comply with the Chinese government's requests for access to their equipment. The companies' responses to the Committee's questions for the record after the hearing included similar evasive answers.

C. Investigative Challenges

This unclassified report discloses the unclassified information the Committee received when trying to understand the nature of these companies, the formal role of the Chinese government or Chinese Communist Party within them, and their current operations in the United States. In pursuing this goal, the Committee faced many challenges, some of which are shared by many who seek to understand the relationship between the government and business in China and the threat posed to our infrastructure. These challenges include: the lack of transparency in Chinese corporate and bureaucratic structures that leads to a lack of trust; general private sector concerns with providing proprietary or confidential information; fears of retribution if private-sector companies or individuals discuss their concerns; and uncertain attribution of cyber attacks.

The classified annex provides significantly more information adding to the Committee's concerns. That information cannot be shared publicly without risking U.S. national security. The unclassified report itself, however, highlights that Huawei and ZTE have failed to assuage the Committee's significant security concerns presented by their continued expansion into the

United States. Indeed, given the companies' repeated failure to answer key questions thoroughly and clearly, or support those answers with credible internal evidence, the national-security concerns about their operations have not been ameliorated. In fact, given their obstructionist behavior, the Committee believes addressing these concerns have become an imperative for the country.

In addition to the Committee's discussions with the companies, the Committee investigators spoke with industry experts and former and present employees about the companies. Companies around the United States have experienced odd or alerting incidents using Huawei or ZTE equipment. Officials with these companies, however, often expressed concern that publicly acknowledging these incidents would be detrimental to their internal investigations and attribution efforts, undermine their ongoing efforts to defend their systems, and also put at risk their ongoing contracts.

Similarly, statements by former or current employees describing flaws in the Huawei or ZTE equipment and other potentially unethical or illegal behavior by Huawei officials were hindered by employees' fears of retribution or retaliation.[32]

Further, the inherent difficulty in attributing the precise individual or entity responsible for known attacks within the United States continues to hinder the technological capability for investigators to determine the source of attacks or any connections among industry, government, and the hacker community within China.[33]

III. Summary of Findings

Chinese telecommunications companies provide an opportunity for the Chinese government to tamper with the United States telecommunications supply chain. That said, understanding the level and means of state influence and control of economic entities in China remains difficult. As Chinese analysts explain, state control or influence of purportedly private-sector entities in China is neither clear nor disclosed.[34] The Chinese government and the Chinese Communist Party, experts explain, can exert influence over the corporate boards and management of private sector companies, either formally through personnel choices, or in more subtle ways.[35] As ZTE's submission to the Committee states, "the degree of possible government influence must vary across a spectrum."[36]

The Committee thus focused primarily on reviewing Huawei's and ZTE's ties to the Chinese state, including support by the Chinese government and state-owned banks, their connections to the Chinese Communist Party, and their work done on behalf of the Chinese military and intelligence services. The Committee also probed the companies' compliance with U.S. laws, such as those protecting intellectual property, to determine whether the companies can be trusted as good corporate actors. The Committee did not attempt a review of all technological vulnerabilities of particular ZTE and Huawei products or components. Of course, the Committee took seriously recent allegations of backdoors, or other unexpected elements in either company's products, as reported previously and during the course of the investigation. But the expertise of the Committee does not lend itself to comprehensive reviews of particular pieces of equipment.

The investigation sought to answer several key questions about the companies that would, including:

- What are the companies' histories and management structures, including any initial ties to the Chinese government, military, or Communist party?
- How and to what extent does the Chinese government or the Chinese Communist Party exert control or influence over the decisions, operations, and strategy of Huawei and ZTE?
- Are Huawei and ZTE treated as national champions or otherwise given unfair or special advantages or financial incentives by the Chinese government?
- What is the presence of each company in the United States market and how much influence does the parent company in Shenzhen influence its operations in the United States?
- Do the companies comply with legal obligations, including those protecting intellectual property rights and international sanctions regimes (such as those with respect to Iran)?

The Committee found the companies' responses to these lines of inquiry inadequate and unclear. Moreover, despite repeated requests, the companies' consistently provided very little – if any – internal documentation substantiating their answers. The few documents provided could rarely be authenticated or validated because of the companies' failure to follow standard document-production standards as provided by the Committee and standard with such investigations. Moreover, the apparent control of the Chinese government over this information remains a hindrance to a thorough investigation. One of the companies asserted clearly both verbally and in writing that it could not provide internal documentation that was not first approved by the Chinese government.[37] The fact that Chinese companies believe that their internal documentation or information remains a "state secret," only heightens concerns about Chinese government control over these firms and their operations.

The Committee is disappointed that Huawei and ZTE neither answered fully nor chose to provide supporting documentation for their claims, especially given that Huawei requested a thorough and complete investigation. The Committee simply cannot rely on the statements of company officials that their equipment's presence in U.S. critical infrastructure does not present a threat, and that the companies are not, or would not be, under pressure by the Chinese government to act in ways contrary to United States interests. The findings below summarize what the Committee has learned from information available, and suggest avenues for further inquiry.

A. The Committee finds that Huawei did not fully cooperate with the investigation and was unwilling to explain its relationship with the Chinese government or Chinese Communist Party, while credible evidence exists that it fails to comply with U.S. laws.

Throughout this investigation, Huawei officials sought to portray the company as transparent. Huawei consistently refused, however, to provide detailed answers in written form or provide internal documentation to support their answers to questions at the heart of the investigation. Specifically, Huawei would not fully describe the history, structure, and management of Huawei and its subsidiaries to the Committee's satisfaction. The Committee received almost no information on the role of Chinese Communist Party Committee within Huawei or specifics about how Huawei interacts in formal channels with the Chinese government. Huawei refused to provide details about its business operations in the United States, failed to disclose details of its dealings with the Chinese military or intelligence services, and would not provide clear answers on the firm's decision-making processes. Huawei also failed to provide any internal documents in response to the Committee's written document requests, thus impeding the Committee's ability to evaluate fully the company's answers or claims.

In addition to discussions with Huawei officials, the Committee has interviewed several current and former employees of Huawei USA, whose employees describe a company that is managed almost completely by the Huawei parent company in China, thus undermining Huawei's claims that its

United States operations are largely independent of parent company. The testimony and evidence of individuals who currently or formerly worked for Huawei in the United States or who have done business with Huawei also brought to light several very serious allegations of illegal behavior that require additional investigation. The Committee will refer these matters to the Executive Branch for potential investigation.

These allegations were not the focus or thrust of the investigation, but they were uncovered in the course of the investigation. The Committee believes that they reveal a potential pattern of unethical and illegal behavior by Huawei officials, allegations that of themselves create serious doubts about whether Huawei can be trusted to operate in the United States in accordance with United States legal requirements and international codes of business conduct.

i. Huawei did not provide clear and complete information on its corporate structure and decision-making processes, and it likely remains dependent on the Chinese government for support.

Huawei markets itself as a "leading global ICT ["Information Communications Technology"] solution provider," that is "committed to providing reliable and secure networks."[38] Throughout the investigation, Huawei consistently denied having any links to the Chinese government and maintains that it is a private, employee-owned company.[39] Many industry analysts, however, have suggested otherwise; many believe, for example, that the founder of Huawei, Ren Zhengfei, was a director of the People's Liberation Army (PLA) Information Engineering Academy, an organization that they believe is associated with 3PLA, China's signals intelligence division, and that his connections to the military continue.[40] Further, many analysts suggest that the Chinese government and military proclaim that Huawei is a "national champion" and provide Huawei market-distorting financial support.[41]

In seeking to understand the Chinese government's influence or control over Chinese telecommunications companies, the Committee focused on Huawei's corporate structure and decision-making processes. Better information about Huawei's corporate structure would also help answer lingering questions caused by Huawei's historic lack of transparency.[42] For years, analysts have struggled to understand how Huawei's purported employee-ownership model works in practice, and how that ownership translates into corporate leadership and decision-making.[43] Huawei repeatedly asserts that it is a private, employee-owned and controlled company that is not influenced by the Chinese government or Chinese Communist Party.[44] Executives also asserted that the unique shareholder and compensation arrangement is the foundation of the company's rise and success.

Available information does not align with Huawei's description of this structure, and many analysts believe that Huawei is not actually controlled by its common shareholders, but actually controlled by an elite subset of its management.[45] The Committee thus requested further information on the structure of the company's ownership. For example, the Committee requested that Huawei list the ten largest shareholders of the company. Huawei refused to answer.[46] At the hearing on September 13, 2012, Huawei admits that its shareholder agreement gives veto power to Ren Zhengfei, the founder and president of the company.[47] Other public statements by the company undermine the suggestion that the 60,000 supposed shareholders of Huawei control the company's decisions. For example, in the company's 2011 report, Mr. Ren highlighted that Huawei's Board of Directors:

> will not make maximizing the interests of stakeholders (including employees, governments, and suppliers) its goal. Rather, it holds on to the core corporate values that are centered on customer interests and encourage employee dedication.[48]

Such statements undermine the credibility of Huawei's repeated claims that its employees control the company. Thus, to explore these conflicts, the Committee focused much attention on the shareholder program. Of note, the only nonpublic, purportedly internal documents that Huawei provided to the Committee in the course of the investigation are unsigned copies of its shareholder agreement documents. Unfortunately, the Committee could not verify the legitimacy of these documents, because they were unsigned and non-official.

Huawei officials explained that Chinese law forbids foreigners from holding shares in Chinese companies absent a special waiver.[49] Current and former Huawei employees confirm that only Chinese nationals working at Huawei in the United States participate in the shareholding plan. The inability of non-Chinese employees of Huawei to hold shares of the company further erodes its claim that it is truly an employee-run organization as an entire group of employees are not only disadvantaged, but automatically excluded from any chance to participate in the process.

Huawei consistently asserted that the Chinese government has no influence over corporate behavior and that the company is instead managed as an employee-owned enterprise through Huawei's Employee Stock Ownership Program (ESOP). Officials explained that the shareholding plan is not a benefits plan; rather, it provides high-performing employees an option to buy dividend-providing shares and thereby share in the value of the company. Eligible employees are given the option to buy shares at a certain company determined price, and can only sell the shares when they leave the company or with approval.[50]

Huawei also provided staff access to shareholder ballots for shareholder representatives and the Board of Directors. These too did not appear to be facially fraudulent, but they were impossible to authenticate, especially as investigators were not allowed to remove the documents from Huawei's facilities for third-party validation. The documents appeared to highlight that shareholders have a write-in option for union representatives, but there is no such option for the Board of Directors. Rather, Huawei officials stated that the nominees for the Board are chosen prior to the vote by the previous Board. It was unclear how the original Board was established, and Huawei has consistently failed to provide any answers about who was previously on its Board of Directors.

Huawei further explained that in 1994, the first Company Law of China was officially published, regulating the establishment and operations of limited liability companies.[51] Under this law, the maximum number of shareholders was 50 individuals. Thus, in 1997, Huawei claims to have changed its legal structure to a limited liability company, and started the employee stock ownership program through the union. Similarly, Huawei asserted that in 1997, the City of Shenzhen issued policies regarding employee shareholdings. According to Huawei, it designed its shareholder program to conform to the the Company Law of China, and the laws and policies of the City of Shenzhen.[52]

According to Huawei, the union, known as Union of Huawei Investment and Holding Co., Ltd., facilitates ESOP implementation. The Union is a lawfully registered association of China. Huawei officials stated that "Huawei's success can be directly linked to the company's unique compensation structure."[53] Currently, Huawei claims that the Union holds 98.7% of the ESOP shares, and Mr. Ren holds 1.3%. At the Huawei explained that as of December 31, 2011, ESOP has 65,596 participants, which it alleges are all Huawei employees (current and retired), it claims that there are no third parties, including government institutions, holding any ownership-stake in the company.

Questions remained after the Committee staff's meeting with Huawei officials. Most importantly, the Committee did not receive clear information about how precisely candidates for the Board of Directors are chosen. This is a concern because such individuals are key decision-makers of the company and those whose potential connections to the government are of high concern. According to Huawei officials, the previous Board nominates the individuals for the current Board. But it is not clear how the original Board was established and Huawei refuses to describe how the first Board of Directors and first Supervisory Board were chosen.[54]

As described above, Huawei provided the Committee unsigned, unauthenticated documents purporting to be: (1) Articles of Restricted Phantom Shares; (2) Letter of Undertakings of Restricted Phantom Shares; (3) Notice of Share Issuance and Confirmation Letter; (4) List of Shareholding Employees; (5) Record of Employee Payments and Buyback, (6) Receipts of Employee Share Payments and Buyback; (7) Election Records of the 2010 ESOP Representatives Election (procedures, ballots, results, announcements, etc.); (8) and conclusions of the 2010 ESOP Representatives Meeting. The Committee could not validate the legitimacy of these documents given that Huawei only provided unsigned drafts. Below are summaries of key information from these documents.[55]

(1) ESOP Restricted Phantom Shares--Summary

- ESOP Restricted Phantom Shares Article 20 states that target grantees of employee stocks are current employees with high performance.
- Each year, the company determines the numbers of shares an employee can purchase based on job performance. Eligible employees must sign the Confirmation Letter and the Letter of Undertakings and make payments for the shares.
- An employee's stocks can be held only by the employee him/herself, and cannot be transferred or disposed by the employee. When an employee leaves the company (except for those who meet the retirement requirements with minimal eight years of tenure and 45 years old), stocks will be purchased back by the company.
- The current stock price is the net asset value of the stock from the previous year. When an employee purchases more shares or the Union takes shares back, it is based on the current stock price. The dividend amount of each year is based on the performance of the company.

(2) Articles of Restricted Phantom Shares[2]

a. The Commission

- The Commission is composed of 51 Representatives and nine alternates, elected by the Active Beneficiaries as organized by the Union with a term of five years.
  - Active beneficiary is defined as an active employee who works at Shenzhen Huawei Investment and Holding Co, Ltd or any of its equity affiliates and participates in the Plan of the Union.
  - In the event there is a vacancy, the Alternate shall take up the vacancy in sequence. The Alternates can attend, but not vote at, all meetings.
  - The Commission reviews and approves restricted phantom share issuance proposals; reviews and approves dividend distribution proposals; reviews and approves reports of the board of shareholding employees; elects and replaces any member of the board; elects and replaces any member of Supervisory Board; reviews and approves procedures for electing representatives; approves amendments of these articles; reviews and approves the use of the reserve fund; reviews and approves other material matters with respect to restricted phantom share; perform functions as the shareholders of the company, exercises the rights of the shareholder, and develops resolutions regarding material matters such as capital increase, profit distribution, and selection of Directors and Supervisors.
- Meetings of the Commission shall be convened at least once a year, and shall be convened by the Board and presided over by the Chairman of the Board or the Vice Chairman.

b. The Board

- The Board is responsible for regular management authority and shall be responsible to the Commission.
- The main functions of the Board are to: prepare restricted phantom share issuance proposal; preparation of the dividends distribution proposal; formulation, approval, and amendment of the detailed rules, processes, and implementation methods with respect to the restricted phantom shares; preparation of the amendments to articles;

determination on the detailed proposal as to the use of the Reserve Fund; execution of the resolutions of the Commission; exercise of the specific rights and powers of a shareholder of the Investee Company except for the matters on which a resolution from Commission is required; determination of other matters that shall be determined by the Board.

- The Board consists of 13 directors selected by the Commission; each serves for five years.
- The Board must convene at least once a year; it needs 2/3 present, and resolutions of the meetings shall be approved by at least 1/2 of all Directors.
- The Board may establish a restricted phantom share management committee and other necessary organizations responsible for carrying out and implementing the work assigned by the Board and for detailed matters with respect to the management of the restricted phantom shares, such as evaluation, distribution, and repurchase of the restricted phantom shares as well as management of the account and the Reserve Fund/treasury shares related to restricted phantom shares.

c. Supervisory Board

- The Supervisory Board is the organization responsible for supervising the implementation of the shareholder plan with its main functions and powers as follows:
    - supervising the implementation of the resolutions by the Board;
    - making recommendations or inquiries in event of any violation of any law, regulation or these Articles by the Board;
    - making work reports to the Commission; and
    - other regular functions and powers.
- Supervisors may attend Board meetings as non-voting delegate.
- The Supervisory Board shall consist of five Supervisors who shall be elected by the Commission to five year terms; no Director can serve concurrently as a Supervisor.
- Convene at least once a year, need minimum of 2/3 present, resolutions require approval of at least 2/3 of all Supervisors

d. Validity of Resolutions

- Before 31 December 2018, Mr. Ren shall have a right to veto the decisions regarding restricted phantom shares and Huawei's material matters (resolutions of the Board, Commission, and Shareholder's Meeting of the Company).
- Starting from 1 January 2013, the confirmed Active Beneficiaries who represent a minimum of 15% of the restricted phantom shares (excluding the restricted phantom shares held by the Restructuring Beneficiaries and the Retained Restricted Phantom Shares) shall have a right to veto the decisions regarding restricted phantom shares and Huawei's material matters (including resolutions of the Board, the Commission, and the Shareholders' Meeting of the Company).
- The relevant resolutions shall take effect in the event that the owner(s) of the right of veto does (do) not exercise the right of veto against the aforementioned resolutions.

(3) Acquisition of Restricted Phantom Shares

- The restricted phantom shares of the Union shall be issued to those key employees of the Company who have displayed excellent work performance.
- The Restricted Phantom Share Management Committee shall decide annually whether to issue shares, and the number of shares to be issued, based on the comprehensive evaluation of the work performance of such employee and in accordance with the evaluation rules of the restricted phantom shares. Retired or restructuring beneficiaries are not allowed to purchase new shares.

(4) Confidentiality and Non-Competition Obligations of the Beneficiaries

- No Active Beneficiary or Restructuring Beneficiary shall directly or indirectly have a second job in any way, work for any enterprise other than the Company without written consent of the Company or without entering into the relevant agreement with the Company.

ii. Huawei failed to explain its relationships with the Chinese government, and its assertions denying support by the Chinese government are not credible.

The nature of the modern Chinese economy is relevant for understanding Huawei's connection to the Chinese state. The Chinese government often provides financial backing to industries and companies of strategic importance. Indeed, analysts of the Chinese political economy state that:

> Huawei operates in what Beijing explicitly refers to as one of seven "strategic sectors." Strategic sectors are those considered as core to the national and security interests of the state. In these sectors, the CCP [Chinese Communist Party] ensures that "national champions" dominate through a combination of market protectionism, cheap loans, tax and subsidy programs, and diplomatic support in the case of offshore markets. Indeed, it is not possible to thrive in one of China's strategic sectors without regime largesse and approval.[56]

Similarly, the U.S.-China Commission has explained, with Chinese companies, "the government's role is not always straightforward or disclosed." Despite some reforms, "much of the Chinese economy remains under the ownership or control of various parts of the Chinese government."[57]

The U.S. China-Commission lists Huawei as a form of enterprise in China that exists in a relatively new market and receives generous government policies to support its development and impose difficulties for foreign competition.[58]

The Committee thus inquired into the precise relationship between the Chinese government and Huawei. During the Committee's meetings with Huawei executives, and during the open hearing on September 13, 2012, Huawei officials consistently denied having any connection to or influence by the Chinese government beyond that which is typical regulation.[59]

Specifically, Huawei explained in its written responses to the Committee, that "Huawei maintains normal commercial communication and interaction with relevant government supervisory agencies, including the Ministry of Industry and Information Technology and the Ministry of Commerce."[60] Huawei claims that it "does not interact with government agencies that are not relevant to its business activities, including the Ministry of National Defense, the Ministry of State Security, and the Central Military Commission."[61] Huawei, however, did not provide information with which the Committee could evaluate these claims, as Huawei refused to answer the specific questions of the Committee inquiring about the company's precise mechanisms of interaction with and regulation by these government bodies.

The Committee did not expect Huawei to prove that it has "no ties" to the government. Rather, in light of even experts' lack of certainty about the state-run capitalist system in China, the Committee sought greater understanding of its actual relationship with the Chinese government.

The Committee requested that Huawei support and prove its statements about its regulatory interaction by providing details and evidence explaining the nature of this formal interaction. Any company operating in the United States could very easily describe and produce evidence of the federal entities with which it must interact, including which government officials are their main points of contact at those regulatory agencies.

In its written submission in response to the Committee's questions, Huawei simply asserted that it "maintains normal commercial communication and interaction with relevant government supervisory agencies, including the Ministry of Industry and Information Technology and the Ministry of Commerce."[62] Huawei's failure to provide further detailed information explaining how it is formally regulated, controlled, or otherwise managed by the Chinese government undermines the company's repeated assertions that it is not inappropriately influenced by the Chinese government. Huawei appears simply unwilling to provide greater details that would explain its relationships with the Chinese government in a way that would alleviate security concerns.

Similarly, Huawei officials did not provide detailed answers about the backgrounds of *previous* Board Members.

Rather, the Committee simply received the same biographies as previously disclosed of current members of the Board of Directors and Supervisory Board.[63]

Previous Board Members may have significant ties to the Party, military, or government. And since the previous Board is responsible for nominating the current Board members, this information is important to understanding the historical progression of the company.

Because the biographies of the previous members would highlight possible connections to military or intelligence elements of the Chinese government, Huawei's consistent failure to provide this information is alerting.

iii. Huawei admits that the Chinese Communist Party maintains a Party Committee within the company, but it failed to explain what that Committee does on behalf of the Party or which individuals compose the Committee.

Huawei's connection to the Chinese Communist Party is a key concern for the Committee because it represents the opportunity for the State to exert its influence over the decisions and operations of a company seeking to expand into the critical infrastructure in the United States. This concern is founded on the ubiquitous nature of the Chinese Party in the affairs of institutions and entities in China, and the consensus view that the Party exerts pressure on and directs the resources of economic actors in China.[64]

In response to the numerous opportunities to answer questions about its connection to the Party, Huawei stated that the company has no relevant connections. For example, in response to the Committee's written questions about the role of the Party in the company's affairs, Huawei merely stated that it "has no relationship with the Chinese Communist Party in its business activities."[65]

Huawei admits, however, that an internal Party Committee exists within Huawei. Huawei simply states that party committees are required by Chinese law to exist in all companies in China.[66] The existence of these Committees is, however, of particular relevance. Huawei states in its defense that all economic institutions in China are required to have a state Party apparatus inside the company. This is not, however, a compelling defense for companies seeking to build critical infrastructure in the United States. Indeed, experts in Chinese political economy agree that it is through these Committees that the Party exerts influence, pressure, and monitoring of corporate activities. In essence, these Committees provide a shadow source of power and influence directing, even in subtle ways, the direction and movement of economic resources in China.[67] It is therefore suspicious that Huawei refuses to discuss or describe that Party Committee's membership. Huawei similarly refuses to explain what decisions of the company are reviewed by the Party Committee, and how individuals are chosen to serve on the Party Committee.

Similarly, Huawei officials did not provide information about Mr. Ren's role or stature in the Party. In his official biography, Mr. Ren admits that he was asked to be a member of the 12th National Congress of the Communist Party of China in 1982. The National Congress is the once-in-a-decade forum through which the next leaders of the Chinese state are chosen. The Party members asked to play a role in China's leadership transition are considered key players in the state apparatus.[68] Mr. Ren proudly admits that he was invited to that Congress, but he will not describe his duties. Shortly after being given such a prestigious role, Mr. Ren successfully founded Huawei, though he asserts he did so without any government or Party assistance.[69] Huawei likewise refuses to answer whether Mr. Ren has been invited to subsequent National Congresses or has played any role in Party functions since that time.[70]

From the review of available information, Huawei may have connections and ties to Chinese leadership that it refuses to disclose. In light of Huawei's refusal to discuss details of its acknowledged Chinese Communist Party Committee, the Committee questions the company's ability to be candid about any other possible connections to the government, military, or Chinese Communist Party.

iv. Huawei's corporate history suggests ties to the Chinese military, and Huawei failed to provide detailed answers to questions about those connections.

Huawei explained the founding and development of the company by focusing on the life and history of Ren Zhengfei, Huawei's founder. According to Huawei officials, Mr. Ren was a member of the Chinese military's engineering corps as a soldier tasked to establish the Liao Yang Chemical Fiber Factory and was promoted as a Deputy Director, which was a professional role equivalent to a Deputy Regimental Chief, but without military rank.[71] Mr. Ren then retired from the army in 1983 after the engineering corps disbanded, and next worked for a State Owned Enterprise (SOE) following his retirement. According to this account, Mr. Ren was "dissatisfied" with his low salary and career path at the SOE, so in 1987, he established Huawei. Huawei officials did not explain how he was able to leave his employment with a SOE or whether he got agreement of the state to do so. Huawei officials denied that Mr. Ren was a senior member of the military.[72] The Committee's requests for more information about Mr. Ren's military and professional background were unanswered. Huawei refused to describe Mr. Ren's full military background. Huawei refused to state to whom he reported when he was in the military. Huawei refused to answer questions about how he was invited to join the 12th National Congress, what duties he performed for the Party, and whether he has been asked to similar state-party matters.

Huawei similarly denied allegations that Ms. Sun Yafang, Chairwoman of Huawei, was previously affiliated with the Ministry of State Security. Mr.

Ding responded to Committee questions after the hearing that, to his knowledge, reports about Ms. Yafang in Chinese publications, such as those in *Xinjing Bao*, are erroneous.[73] Mr. Ding did not respond to questions asking about how such publications received such information, or whether Ms. Yafang's previous biography on the Huawei website was erroneous as well. Rather, Mr. Ding simply provided again Ms. Yafang's corporate biography from the Huawei Annual Report 2011.[74]

With respect to Huawei's founders, Huawei cited a Chinese legal requirement that new companies in the economic development zone must have a minimum of five shareholders and 20,000 RMB registered capital. During meetings with the Committee, Huawei officials claimed that in 1987, Mr. Ren raised 21,000 RMB with personal savings and five other private investors. To the best of the officials' knowledge, none of the five investors had worked with Mr. Ren prior to start-up and one individual has previous affiliation with the government.[75] According to Huawei officials, the five investors never actually worked for Huawei and withdrew their investments several years later.[76]

The Committee struggled to get answers from Huawei on the details of this founding, including how Mr. Ren came to know the initial individual investors, whether his connections to the military were important to the eventual development of the firm, and whether his role in the Party remains a factor in his and his company's success.

v. The Committee finds that Huawei's failure to provide information about the Chinese government's 1999 investigation of the company for tax fraud exemplifies how it refuses to be transparent; the apparent ease with which Huawei ended the investigation undermines Huawei's assertion that the Chinese government finds Huawei to be a disfavored telecommunications solutions provider in China.

Huawei officials claimed that after growing in rural areas in China throughout the 1990s, the Chinese government investigated the company at length between 1998-99 for tax fraud.[77] Huawei officials stated that they believed this investigation was politically motivated and performed at the urging of the company's competition – specifically, other telecommunications companies that are also state-owned enterprises. Mr. Ken Hu explained the investigation was a turning point in the history of the company. Specifically, Mr. Hu stated that Huawei's movement to opportunities outside of China was the result of this investigation.[78] Indeed, these officials sought to explain that this episode proves that Huawei was not in fact a "national champion" or otherwise a favored company in China.[79]

Given the obvious importance Huawei placed on this tax-fraud investigation, the Committee's subsequent questions and document requests sought detailed information and further documentary support for its version of events. In particular, the Committee sought information on the conclusion of the Chinese investigation. This information is particularly important to the Committee given the apparent pride displayed by certain Huawei officials in Shenzhen when describing how they successfully used their connections to end the investigation. The ability of these corporate officers to end a politically-motivated investigation suggests that Huawei officials were not as lacking in political power or influence with the government as they suggested.

Despite the importance placed on this event, Huawei failed to address the Committee's questions in its written submission.[80] The company also failed to provide any material that would support Huawei's assertions that the investigation was closed legitimately or without attendant conditions placed on Huawei.[81]

vi. Huawei failed to explain its relationships with western consulting firms, and any claims that its success is on account of those relationships, rather than support by the Chinese government, are not credible.

Huawei officials stated that one reason for the company's success was its reliance on the advice of western consulting firms, such as IBM, Accenture, and Price Waterhouse Cooper.[82] Huawei sought to convince the Committee that it was the advice of these companies -- and not support by the Chinese government -- that explains Huawei's miraculous growth in recent years.[83]

Because of the importance Huawei places on the advice given by these consulting firms, the Committee sought greater information and evidence showing that such advice had important effects for the company. The Committee made clear that it did not seek information on the terms of the contractual arrangements with the consulting firms, but rather what information they reviewed from Huawei and what advice was provided. The Committee offered to keep such information confidential to avoid concerns about disclosing proprietary information.

Huawei responded with only a vague description of the advice provided by these companies. Specifically, although "[s]ince 1997, Huawei has relied on western management consulting firms to help improve [its] capabilities, build [its] processes, and develop a comprehensive management system driven by customer requirements," Huawei failed to provide details about how those companies reformed the company other than providing a few sentences mentioning standard business practices, including lead to cash (LTC), integrated product development (IPD), issue to resolution (ITR), and integrated financial services (IFS). Huawei, refused "to provide additional details as to [its] consultancy relationships" citing concerns about proprietary information contained in that advice.[84] The Committee explained that it is most interested in evidence revealing what Huawei did in response to the advice of these firms, and specifically financial or other evidence that supports its position that those changes were responsible for efficiencies, growth, and market success.[85] Huawei could have answered such questions without revealing proprietary information held by these companies.[86] The Committee was also willing to enter into a confidentiality agreement with all parties, an offer Huawei declined to accept or pursue.[87]

Huawei has made the details of this consulting advice relevant to this investigation by attributing its rapid success to the advice rendered by these consulting firms. It is not then reasonable for Huawei to withhold that information from the Committee so that it could evaluate those claims. If Huawei has within its possession information and documents that would prove that the advice given by these firms was key to Huawei's success, Huawei should provide such information.[88] The Committee was and remains willing to enter into confidentiality agreements with all parties to solve any concerns about the release of proprietary information. Huawei has failed to accept this offer. Its failure to do so is indicative of the lack of cooperation shown throughout this investigation.

vii. Huawei failed to answer key questions or provide supporting documentation for its claims to be financially independent of the Chinese government.

As a company of strategic importance to China, Huawei's stature will be reflected in the level of financial support and direction it receives from the Chinese government and Party.[89] One way to review that support and direction provided by the state is through the financing the company receives. Many industry experts and telecommunications companies describe below-market pricing.[90] Thus, the Committee sought more information about Huawei's financing, including its customer financing. Such financial information would also help provide greater understanding about the financial structuring of a firm that remains largely opaque.

During the Committee's hearing, Mr. Ding suggested he did not understand and had no knowledge of the term "national champion," which is often used to describe favored Chinese companies throughout the economic literature on China.[91] The Committee finds that Mr. Ding's suggestion that he does not understand the term is not credible. Huawei itself provided Capitol Hill offices a slide presentation in November 2011, which used the term "national champion" several times.[92] In response to the Committee's questions about use of the term in that document, Huawei did not deny that it used the document and provided the document containing the term.[93] Rather, Huawei stated that the particular slide in the larger document was created by a third party and thus not Huawei's responsibility.[94] The Committee finds that Huawei's knowing use of the document in its discussions with United States elected representatives is sufficient evidence to prove that Huawei does in fact have an understanding of the term. Mr. Ding's consistent refusal to answer questions about which firms are considered national champions in the Chinese telecommunications sector was obstructionist. In fact, his response to the Committee's question that "Huawei has not paid attention to the meaning of 'national champion' before," is obviously untrue given the company's use of the term in its presentations previously.[95] Moreover, his answers suggest that he did not want to explain how it was that Huawei, the number one telecommunications provider in China, is not a company of strategic importance in China, as recognized by others around the world.

Huawei officials also deny that they have received any special financial incentives or support from the Chinese government.[96] Huawei claimed that the company simply takes advantage of general Chinese banking opportunities, but does not seek to influence or coordinate with banks such as the Chinese Development Bank and the Export-Import Bank, which are both state owned. In previous presentations, Huawei had suggested that it served as an "intermediary and bridge" between the state-backed financial institutions and Huawei customers.[97] Huawei refused, however, to provide more detail about precisely how those lines of credit developed. Huawei also refused to answer specifics about its formal relationships with the Chinese banks, opting to simply answer that it maintains "normal business relations" with the Export-Import Bank of China.[98]

In its presentation to the Committee during the February meeting, Huawei provided a list of the Memoranda of Understanding (MOUs) it claims to have signed with Chinese banks for lines of credit for its customers.[99] Huawei admits that its customers have a US $100 billion in credit available, yet Huawei asserts that only $5.867 billion has been drawn in the period between

2005 and 2011. Further, in its written responses, Huawei states that it is a "financing opportunity available to customers, not to Huawei."[100] Yet Huawei explained at the February 23, 2012, meeting with Committee investigators that the goal of the large available credit lines was for China "to appear impressive" and that "Huawei had to participate or would no longer receive loans" from Chinese banks.[101] In response to repeated questions and document requests, Huawei failed to provide further written explanation of the benefits Huawei gains from these financing arrangements, and it did not provide internal documents or any auditable information that would substantiate its claims about the scope and processes of this financing.

Similarly, Huawei refused to describe the details of its relationships with Chinese state-owned banks. For example, in Mr. Ding's statement for the record, he explained that Huawei receives loans from ten Chinese banks. But Mr. Ding refused to answer how many of those ten banking institutions in China are state-owned.[102] As described in the previous section, Huawei also refused to provide "additional details as to [its] consultancy relationships" because it would "include highly sensitive proprietary information" governed by non-disclosure agreements.[103] In response to Committee questions about Huawei's success and whether it was owing to the company's support from the Chinese government, Huawei represented to the Committee that its relationships with and advice received from these companies are the source of the company's global success.[104] Because Huawei refuses to provide details on those relationships and advice rendered, the Committee cannot evaluate its claim that any of its success is due to these relationships. Accordingly, the Committee discounts the role played by these consulting companies, and continues to find it likely that Huawei has substantially benefited from the support of the Chinese government.

In sum, Huawei admits that its customers receive billions of dollars in support from Chinese state-owned banks and that it has received favorable loans from Chinese banks for years. Huawei refuses to provide answers to direct questions about how this support was secured, nor does it provide internal documentation or auditable financial records to evaluate its claims that the terms of these agreements comply with standard practice and international trade agreements. The Committee is equally concerned with statements by company leaders that undermine the Committee's confidence in the financial information the company has provided. For example, in a June 2007 speech to Huawei employees in the United Kingdom, Mr. Ren stated that he appreciated the subsidiary's attempt to create financial statements, "whether the data is accurate or not."[105] Based on available information, the Committee finds that

Huawei receives substantial support from the Chinese government and Chinese state-owned banks, which is at least partially responsible for its position in the global marketplace.

viii. Huawei failed to provide sufficient details or supporting documentation on its operations, financing, and management in the United States; evidence undermines its claims of being a completely independent subsidiary of Huawei's parent company in Shenzhen, China.

To understand the United States' current vulnerability to supply-chain threats posed by Huawei equipment, it is necessary to know the extent to which Huawei's equipment is already placed in U.S. infrastructure. Because the U.S. telecommunications infrastructure is largely built and owned by the private sector, the U.S. government does not have the full picture of what is contained within it and thus is not yet fully informed to develop policies to protect that critical infrastructure from vulnerabilities.[106]

The Committee thus asked Huawei for information on its contracts for products and services within the United States. Understanding the extent to which Huawei equipment already exists in the United States is necessary to evaluate the present risk to the country, as well as to confirm Huawei's statements about the size and scope of its operations in the United States. Unfortunately, Huawei failed to provide specific information about its dealings within the United States. Huawei did provide the Committee a list of Huawei's major customers in the United States: Cricket Communications; Clearwire; Cox TMI Wireless, Hibernia Atlantic, Level 3/BTW Equipment, Suddenlink; Comcast and Bend Broadband. Huawei, however, did not provide information on the size and scope of its operations, which elements of the infrastructure it is providing, and where these operations are located.[107]

The information requested by the Committee about Huawei's contracts in the United States is also necessary to evaluate Huawei's claims that they comply with all laws and trade obligations regarding the price of their products and services.[108] To date, Huawei has failed to provide any information to validate its claims that the prices of Huawei's products are based on market conditions. Huawei's refusal to answer clearly or provide documents supporting its claims necessitates the Committee finding that Huawei's defense is not credible. The Committee considers it possible that Huawei receives substantial support from the Chinese government such that Huawei can market at least some of its products in the United States below the costs of production.

Similarly, the extent to which Huawei's subsidiaries in the United States operate independently of the parent company in Shenzhen remains unclear. Such information is important, because any connections between the parent company in China to the Chinese government might affect the operations and behavior of the company in the United States. The Committee therefore requested information on the extent to which Huawei USA's decisions are controlled, influenced, or reviewed by the parent company.

Huawei explained that the first US-based Huawei subsidiary was established in the United States in 2005 with headquarters in Plano, Texas. Huawei stated that the parent company does not require approval for individual contracts in the United States.[109] Rather, it stated that the Board of Directors in China does set general terms for operations in the United States, and the parent company can help mobilize resources and set strategy should the subsidiary need it. The Committee has heard from several former Huawei employees in the United States who dispute Huawei's explanation of its business model. Sources from around the United States have provided numerous specific instances of business decisions in the United States requiring approval by the parent company in China. In one instance, an individual with first-hand knowledge explained that senior level executives in the United States could not sign a contract for cyber-security services in the United States without approval in China. In fact, in one instance, a contract previously signed by a U.S.-based senior official at Huawei was repudiated by the parent company.[110] These employees provided documentary evidence, including internal memoranda and emails, discussing corporate policy from China. This description of Huawei's US subsidiaries also comports with reports about the ties between other Huawei subsidiaries and the parent company in China.[111]

To resolve this conflict, the Committee sought more information through its written questions to understand the precise mechanisms through which the Huawei parent company in Shenzhen controls Huawei's strategy for entry and growth into the United States market. Concerns that Beijing's support to Huawei could impact the U.S. market were heightened by Huawei officials' statements to Committee staff that Huawei USA is given general guidance and "resources" from the parent company if needed.[112] In its written response, however, Huawei failed to answer the Committee's detailed questions or provide any further information about the level of coordination between Huawei USA and the parent company.[113]

The information and material provided by Huawei employees with first-hand access coupled with Huawei's failure to provide detailed, internal information, undermines Huawei's claims. For these reasons, the Committee does not find credible Huawei's claims that its U.S. subsidiaries operate independently of the Huawei headquarters in Shenzhen, China.

ix. Evidence shows that Huawei exhibits a pattern of disregard for the intellectual property rights of other entities and companies in the United States.

Huawei's ability to protect intellectual property rights is an important indicator of the company's ability to abide by the laws of the United States. Thus, the Committee sought greater information on Huawei's history of IP protection.

The Committee has reason to believe that Huawei is careless with its compliance with intellectual property protections. Investigators heard from numerous sources that Huawei has a checkered history when it comes to protecting the intellectual property of other entities.[114] Specifically, several former employees of Huawei said it is known to purposely use the patented material of other firms. First-hand accounts of former employees suggest that Huawei does not appropriately purchase software applications for use by its employees.[115] Similarly, the Committee heard from industry experts that Huawei has purposely used and marketed patented products of other companies.[116] Finally, the Committee is in receipt of a Huawei slide presentation that was provided to Capitol Hill offices that itself violates copyright obligations by knowingly using proprietary material from an outside, nonaffiliated consulting firm.[117]

Huawei officials consistently denied ever infringing other companies' intellectual property rights. Even with respect to the litigation with Cisco, in which Huawei agreed to remove certain products from the marketplace, Huawei asserts that it had not violated Cisco's interests.[118] Rather, Huawei suggested that the expert's review in that case of their equipment found no infringement of Cisco patents.[119]

Huawei's defense is not credible. First, with respect to the Cisco litigation, Huawei's statements do not comport with statements made by Huawei officials at the time of the lawsuit acknowledging that the company will remove infringing equipment.[120] Second, the Cisco settlement itself required Huawei to "update and change all of the products that have been accused of violating copyright or intellectual property rights."[121] Finally, during the hearing on September 13, 2012, Charles Ding refused to answer the clear question of whether Cisco code had ever been in Huawei equipment.[122] Mr.

Ding's obstructionism during the hearing undermines Huawei's claims that it did not violate Cisco's patented material.

The Committee finds that Huawei's denials of intellectual property infringement were not credible or supported by available evidence. Because Huawei failed to produce any internal documents or support for its defenses, the Committee finds that Huawei has exhibited a pattern of, at the very least, reckless disregard for the intellectual property rights of other entities.

x. Huawei failed to provide details of its operations in Iran, though it denied doing business with the government of Iran, and did not provide evidence to support its claims that it complies with all international sanctions or U.S. export laws.

Huawei's ability to comply with international sanctions regimes and U.S. export control regulations is an important indicator of the company's ability to comply with international standards of corporate behavior and to abide by U.S. laws irrespective of China's influence or interests. Public reporting raises questions about the company's compliance with these laws.

In response to the Committee's questions, Huawei officials provided only vague assertions about their commitment to all laws. Specifically, Huawei asserted that the company seeks to abide by all legal obligations and has transformed its business practices with the help of outside consultants to better monitor its actions to ensure compliance with international sanctions regimes. To highlight the lack of influence of the Chinese regime over its decisions, Huawei indicated the Chinese Embassy in Iran was surprised by Huawei's decision to limit its business dealings in Iran. Huawei also stated that it does not allow its employees to participate in cyber activities, such as population monitoring, anywhere in Iran.

Huawei has refused, however, to answer detailed questions about its operations in Iran or other sanctioned countries. In its written submission to the Committee, Huawei again reiterated that it limited its future business in Iran because of the enhanced sanctions and an inability to collect payment for operations in Iran. Huawei highlights, though, that "Huawei respects the contracts signed with [its] customers" and thus will not end current contracts in Iran.[123] Huawei claims to "observe laws and regulations of the UN, the U.S., the E.U. and other countries and regions on sanctions."[124] It also claims to have instituted an internal program on trade compliance representing best practices to manage these issues.[125] But Huawei refused to provide any internal documents relating to its decision to scale-back operations in Iran or otherwise ensure compliance with U.S. laws.

Such documents would have validated Huawei's claims that the decision was based on legal compliance requirements and not influenced by any pressure by the Chinese government.

xi. Huawei refused to provide details on its R&D programs, and other documents undermine its claim that Huawei provides no R&D for the Chinese military or intelligence services.

To understand the extent to which Huawei performs R&D activity on behalf of the Chinese military or intelligence services, the Committee asked for information about its activities on behalf of the Chinese government or military. Specifically, the Committee requested information on the technologies, equipment, or capabilities that the funding or grants by the Chinese government was supporting. In its written submission to the Committee, Huawei failed to provide responsive answers to the Committee's questions about the specifics of government-backed R&D activities.[126] Rather, Huawei simply asserted that it only bid on R&D projects open to the rest of the industry.[127]

Huawei similarly claimed in its meetings with the Committee that it does not provide special services to the Chinese military or state security services.[128]

In its answers to the Committee's questions after the hearing, Huawei again simply asserted that it "has never managed any of the PLA's networks" and "has never been financed by the Chinese government for R&D projects for military systems." Huawei did admit, however, that it develops "transport network products, data products, videoconferencing products, and all centers, and voice over IP (VoIP) products" for the Chinese military "accounting for .1% of Huawei's total sales."[129] Huawei also claimed, however, that it "develops, researches, and manufactures communications equipment for civilian purposes only."[130]

The Committee also received internal Huawei documentation from former Huawei employees showing that Huawei provides special network services to an entity the employee believes to be an elite cyber-warfare unit within the PLA.[131] The documents appear authentic and official Huawei material, and the former employee stated that he received the material as a Huawei employee.[132] These documents suggest once again that Huawei officials may not have been forthcoming when describing the company's R&D or other activities on behalf of the PLA.

The Committee finds that Huawei's statements about its sales to the Chinese military are inherently contradictory. The Committee also finds that Huawei's failure to respond fully to questions about the details of its R&D activities, coupled with its admission that it provides products for the Chinese military and documents received from employees, undermine the credibility of its assertion that it does not perform R&D activities for the Chinese government or military.

xii. Former and current Huawei employees provided evidence of a pattern and practice of potentially illegal behavior by Huawei officials.

During the course of the investigation, several former and current Huawei employees came forward to provide statements and allegations about Huawei's activities in the United States. Given the sensitivities involved, and to protect these witnesses from retaliation or dismissal, the Committee has decided to keep the identities of these individuals confidential. The Committee has received multiple, credible reports from these individuals of several potential violations by Huawei officials. Those allegations include:

- Immigration violations;
- Bribery and corruption;
- Discriminatory behavior; and
- Copyright infringement.

Specifically, the Committee heard from numerous employees that Huawei employees visiting from China on tourist or conference visas are actually working full-time at Huawei facilities, in violation of U.S. immigration law. Similarly, Huawei employees provided credible evidence that individuals coming to the United States on particular visas, for example, for jobs requiring engineering expertise were not in fact employed by Huawei as engineers. These and other alleged violations of immigration law will be referred to the Department of Homeland Security for review and potential investigation.

Second, employees have alleged instances fraud and bribery when seeking contracts in the United States.[133] Those allegations will be referred to the Justice Department for further review and potential investigation.

Third, employees with whom the Committee spoke discussed allegations of widespread discriminatory behavior by Huawei officials. These individuals assert that it is it very difficult if not impossible for any non-Chinese national to be promoted in Huawei offices in the United States. Further, these employees assert that non-Chinese nationals are often laid-off only to be replaced by individuals on short-term visas from China.[134] These allegations will be referred to the appropriate agencies in the Executive Branch to review and consider.

Finally, the Committee heard from former Huawei employees that may constitute a pattern and practice of Huawei using pirated software in its United States facilities. As previously described, the Committee received information with Huawei's logo that knowingly and admittedly violated another firm's copyrighted material.[135] The Committee thus finds that Huawei has exhibited a careless disregard for the copyrighted material of other entities. As there may indeed be credibility to these employee allegations, the Committee will also refer these claims to the Justice Department for investigation.

B. ZTE failed to answer key questions or provide supporting documentation supporting its assertions, arguing that answering the Committee's requests about its internal corporate activities would leave the company criminally liable under China's states-secrets laws.

ZTE sought to appear cooperative and timely with its submissions to the Committee throughout the investigation. ZTE consistently refused, however, to provide specific answers to specific questions, nor did the company provide internal documentation that would substantiate its many claims. As with Huawei, the Committee focused its review of ZTE on the company's ties to the Chinese state, as well as the company's history, management, financing, R&D, and corporate structure. The Committee did not to receive detailed answers on a number of topics described below. ZTE did not describe its formal interactions with the Chinese government. It did not provide financial information beyond that which is publicly available. It did not discuss the formal role of the ZTE Communist Party Committee and only recently provided any information on the individuals on the Committee. The Committee did not receive details on ZTE's operations and activities in Iran and other sanctioned countries. Finally, ZTE refused to provide detailed information on its operations and activities in the United States.

Importantly as well, ZTE argued at great length that it could not provide internal documentation or many answers to Committee questions given fear that the company would be in violation of China's state-secrets laws and thus subject ZTE officials to criminal prosecution in China.[136] In fact, ZTE refused even to provide the slides shown to the Committee staff during the meeting in April, 2012, for fear that they might be covered by state secrets. To the extent ZTE cannot provide detailed and supported answers to the Committee because China's laws treat such information important to the security of the Chinese regime, the Committee's core concern that ZTE's presence in the U.S. infrastructure represents a national-security concern is enhanced.

The Committee notes that ZTE's many written submissions were never numbered to align with the Committee's specific questions and document requests, as would be typical with formal legal processes. The Committee believes that, through this method, ZTE sought to avoid being candid and obvious about which questions it refused to answer. Moreover, ZTE's answers were often repetitive, lacking in documentary or other evidentiary support, or otherwise incomplete.

The Committee also notes that ZTE did not simply deny all national-security concerns arising from the global telecommunications supply chain. ZTE has advocated for a solution – one based on a trusted delivery model – in which approved "independent third-party assessors" transfer "hardware, software, firmware, and other structural elements in the equipment to the assessor."[137] Such a model, as advocated by ZTE, would include among other things, a "thorough review and analysis of software codes," "vulnerability scans and penetration test," "review of hardware design and audit of schematic system diagram," "physical facility review and independent comprehensive audit of vendor's manufacturing, warehousing, processing, and delivery operations," "periodic assessments."

ZTE suggests that a model, as previously proposed by Huawei and other companies, and similar in some respects to that introduced in United Kingdom, be implemented across the spectrum for telecommunications equipment providers. As discussed above, the Committee remains concerned that, although mitigation measures can be of some assistance, this model fails to appreciate the nature of telecommunications equipment.

i. ZTE did not alleviate Committee concerns about the control of Chinese state-owned enterprises in ZTE's business decisions and operations.

As with Huawei, the Committee is concerned with the influence of the Chinese state in ZTE's operations. Such access or influence would provide a ready means for the Chinese government to exploit the telecommunications infrastructure containing ZTE equipment for its own purposes. To evaluate the ties to the Chinese state, the Committee focused on the company's history, structure, and management. Many commentators have noted that ZTE's founding included significant investment and involvement by Chinese state-owned enterprises, and thus the Committee sought to unpack the current operations and ownership structure with the hope of understanding whether there are remaining ties to the Chinese state.

ZTE describes itself as a global provider of telecommunications equipment and network solutions across 140 countries. Founded in 1985, ZTE states that its 2011 revenue led the industry with a 24% increase to $13.7 billion; its overseas operating revenue grew 30% to U.S. $7.42 billion during the period, accounting for 54.2% of overall operating revenue.[138] ZTE markets itself by explaining that its systems and equipment have been used by top operators in markets around the world. Importantly, ZTE also highlighted in its 2011 Annual Report that China's 12th five-year national plan has significantly contributed to ZTE's recent domestic success.[139]

During the interviews with ZTE officials in April and May 2012, ZTE officials stressed that ZTE is a publicly traded company, having been listed on the Shenzhen stock exchange in 1997, and the Hong Kong stock exchange in 2004. ZTE contends that it did not begin with government assistance, either with technology transfers or financial assistance. Rather, ZTE stated that the Chinese government became a shareholder during the 1997 public offering. ZTE has also asserted that the state-owned-enterprise shareholders have no influence on strategic direction of the company.[140] ZTE officials often contrasted themselves with Huawei, though often did not mention Huawei by name. In particular, officials suggested that Huawei is ZTE's main competitor, but often stated that ZTE is more transparent since it is a publicly traded company.

These officials often relied on this public listing to claim that ZTE finances are transparent and comply with both Chinese and Hong Kong regulations regarding financial disclosures. The officials often simply referred to the fact that they have annual reports that detail information requested, such as amount and extent of government loans, subsidies, and credits. ZTE refused, however, to explain whether it would be willing to meet the reporting and transparency requirements of a western stock exchange such as the New York Stock Exchange.[141] As with Huawei, when the Committee sought more detailed answers from ZTE on its interactions with key government agencies, ZTE refused to answer.

The history and structure of ZTE, as admitted by the company in its submissions to the Committee, reveal a company that has current and historical ties to the Chinese government and key military research institutes. In response to questioning, the ZTE officials first discounted and seemingly contradicted their own public statements, which suggest that ZTE formed originally from the Ministry of Aerospace, a government agency. In fact, exhibits displayed during the meeting in Shenzhen highlighted an early collaboration between ZTE and the government-run No. 691 Factory, and other state-owned enterprises. ZTE refused to provide the Committee copies of the slides presented during this meeting.

ZTE officials instead suggested that Mr. Hou Weigui founded ZTE in 1985 with five other "pioneer" engineers. Although they had all previously worked for state owned enterprises, ZTE officials insisted that the formation of ZTE did not arise from any relationship with the government. The company's written submission to the Committee admits that the company had an early connection to No. 691 Factory, which was established by the Chinese government.[142] As described by ZTE, No. 691 Factory is now known as Xi'an Microelectronics Company, and is a subsidiary of China Aerospace Electronics Technology Research Institute, a state-owned research institute. In its submission, ZTE admits that Xi'an Microelectronics owns 34% of Zhongxingxin, a shareholder of ZTE.[143] ZTE's evolution from research entities with connections to the Chinese government and military highlight the nature of the information-technology (IT) sector in China. In fact, taking as true ZTE's submission of its history and ownership, ZTE's evolution confirms the suspicions of analysts who study the IT sector in China and describe it as a hybrid serving both commercial and military needs.[144]

In 1997, ZTE was publicly listed for the first time on the Shenzhen stock exchange. ZTE executives claim that it was at this point that other state owned enterprises began investing in ZTE.

Currently, 30% of ZTE is owned by Zhongxinxin group and the remaining 70% is held by dispersed public shareholders. The Committee is particularly interested in whether the 30% ownership by Zhongxingxin constitutes a controlling interest or otherwise provides the state owned enterprises an opportunity to exert influence over the company. This question is particularly relevant because two state owned enterprises own 51% of Zhongxingxin. ZTE executives stressed that the public ownership of ZTE is increasing as Zhongxingxin sells its shares (for example, in 2004, Zhongxingxin owned 44% of ZTE, and now Zhongxingxin holds only 30%). In ZTE's July 3 submission to the Committee, ZTE states that "[v]ery few knowledgeable individuals in China would characterize ZTE as a state-owned entity (SOE) or a government-controlled company."[145] But the Committee specifically asked how it is that ZTE remains accountable to its shareholders and not influenced or controlled by its largest shareholder given this ownership structure. In its submission to the Committee, ZTE admits that a 30% share is the point at which Hong Kong and Chinese law considers the holder to be a "controlling shareholder."[146] ZTE simply stated that the company's fiduciary duty to the numerous shareholders means that the controlling shareholder does not in fact exert much actual control over the company.[147] ZTE does not explain in more detail how the Board members, five of whom are chosen by state-owned enterprises, and some of whom are acknowledged members of the Chinese Communist Party and members of ZTE's internal Communist Party Committee, would not exert any influence over the decisions of the company.

Zhongxingxin, ZTE's largest shareholder is owned in part by two state-owned enterprises, Xi'an Microelectronics and Aerospace Guangyu, both of which not only have ownership ties to the Chinese government, but also allegedly partake in sensitive technological research and development for the Chinese government and military. ZTE failed to address the Committee's questions seeking detailed information on the history and mission of these two companies. ZTE also failed to answer questions about these companies' relationship to key leaders within ZTE, specifically Mr. Weigu Hou, and ZTE's other major shareholder, Zhongzing WXT.

Because ZTE failed to answer key questions about its history and the connections to government institutions, the Committee cannot trust that it is free of state influence, particularly through its major shareholders and Board members.

ii. ZTE maintains a Chinese Party Committee within the company, and has not fully clarified how that Committee functions, who chooses its members, and what relationship it has with the larger Chinese Communist Party.

As with Huawei, ZTE's connection to the Chinese Communist Party is a key concern for the Committee. Such a connection offers the Party the opportunity to influence the decisions and operations of a company seeking to expand into the critical infrastructure in the United States. As described previously, the modern Chinese state-capitalistic economy is largely influenced if not controlled by the Party, in large part through the party committees that exist within individual firms.

During interviews with ZTE officials, ZTE refused to answer whether the executives or board members are members of the Chinese Communist Party.

ZTE first downplayed the existence of the Party Committee within ZTE, and company representatives were unable to answer whether any members of the Board were also members of the state Party. Subsequently, in response to continued Committee questions, ZTE acknowledged that it does, in fact, contain an internal Committee Party, which ZTE suggests is required by the laws of China.[148] In response to the Committee's written questions, ZTE again refused, however, to provide information about the names and duties of the Party Committee members. At the September 13, 2012, hearing, Mr. Zhu attested under oath that ZTE would provide the names of those individuals on the Party Committee.[149]

In response to questions posed at the September 13, 2012, hearing ZTE did provide the Committee a list of 19 individuals who serve on the Communist Party Committee within ZTE. At least two of those individuals appear to be on the ZTE Board of Directors. Other individuals are major shareholders in ZTE entities. ZTE has requested and the Committee has agreed to keep the names of these individuals out of the public domain. ZTE discounts the influence of that these individuals may have over the company. The company asked that the Committee not release the names of the individuals for fear that the company or the individuals might face retaliation by the Chinese government or Communist Party. The Committee has decided to keep the names of those members out of this public report, but the company's concern with the potential retaliatory measures it faces by the government for simply providing the Committee the names of an internal ZTE body highlights why this Committee remains very concerned that the Chinese state is, or could be, responsible for the actions of the company. China clearly seeks to maintain deep ties into and secrecy about its role in economic actors in China. The control Chinese government maintains over the company's actions and level of transparency is of particular concern when that company seeks to build critical U.S. infrastructure.

ZTE also did not fully explain the function of the Party Committee. Instead, ZTE simply states that the Party Committee is controlled by company management. This assertion is contradicted by ZTE's own statement that ZTE executives and board members actually are members of the [Chinese Communist Party] and delimit its activities."[150] To the extent these executives and Board Members have obligations to both the company's shareholders and the State through the Communist Party, there is an inherent conflict of interest in their duties, and this statement provides confirmation that the Party likely does in fact have influence and input into the business affairs of the company through these individuals.

The affidavit by the independent director, Timothy Steinert, seeks to allay any concerns of influence by the government or Party by stating that:

> In my experience and to my knowledge, no member of ZTE's Board of Directors has raised for consideration an interest on behalf of the Chinese Government, the People's Liberation Army or the Chinese Communist Party.[151]

This statement does not resolve the Committee's concerns. First, there is a range of operational and strategic decisions made on a daily basis within companies that are not decided by or reviewed by the Board. Mr. Steinert's affidavit says nothing about the role of the Party Committee in those decisions prior to their reaching the Board, or for decisions that do not reach the Board at all. Second, the Party's influence through ZTE's Party Committee may not be facially obvious in the decision documents appearing for review to the Board. Since at least two members of the Board are also members of the Chinese State Party, it is impossible to know whether the votes of the Board are conducted without influence by the Chinese Communist Party. When considering ZTE's activities or voting on certain measures, those Board members need not cite the Party to be acting on the state's behalf or in pursuit of the state's interests. For these reasons, the Committee finds unpersuasive ZTE's claims that Mr. Steinert's affidavit "confirms that ZTE business decision making is not influence by the government or Party considerations"[152]

ZTE recently suggested that the Party Committee "performs only ceremonial and social functions." For six months, the Committee has asked ZTE about the role of the Party Committee, but only at the final hour, did it provide any response at all. Without further information and specifics about the role and influence of the Party Committee in the operations of the company, the Committee simply cannot allay the concerns about the internal party apparatus existing within a company seeking to build U.S. critical infrastructure.

iii. ZTE failed to disclose information about its activities in the United States.

ZTE discussed its extensive presence in 140 countries, but significantly downplayed any potential threats to the U.S., by suggesting that 95% of its U.S. sales are from handsets. ZTE officials highlighted that they have five R&D centers in the U.S. employing about 300 people. ZTE officials attempted to suggest that the company's presence in rural infrastructure and networks was to assist the U.S. effort with its rural broadband plans. Committee staff questioned this logic, and ZTE officials admitted that ZTE's role in these projects were not for charity or public service, as they had initially suggested, but to get a "foothold" in the country and learn the technology in the United States. ZTE officials even admitted that they are willing to provide this equipment to the U.S. below cost in order to learn the U.S. market. Specifically, during the Committee's meeting with ZTE officials in Shenzhen, Mr. Zhu stated that the company was willing to lose money on projects in the United States to get a foothold in the United States and to understand the technology and standards in the United States.

ZTE's description of its current U.S. activity is simply a picture at a particular point in time. The Committee could not confirm the extent of the company's contracts or access to the United States market absent responses to the Committee's document requests.[153] Despite numerous requests, ZTE has not provided detailed information on infrastructure projects in the United States.[154] ZTE also failed to answer follow-up questions that would explain whether ZTE purposely bids on projects below cost and how the company is able to sustain these losses. Further, at the HPSCI hearing on September 13, Mr. Zhu reversed his previous answers and refused to acknowledge that ZTE ever bids below cost for projects in the United States.[155]

iv. ZTE failed to provide any answers or evidence about its compliance with intellectual property or U.S. export-control laws.

The protection of intellectual property and compliance with United States export control laws are a core concern for U.S. interests. The ability of a company to comply with these laws provide a useful test of that company's ability to follow international codes of business conduct and remain free of undue state influence.

Representatives of the company consistently declined to comment on recent media reports that ZTE had sold export-controlled items to Iran.[156] At the hearing on September 13, 2012, ZTE acknowledged that it is performing an internal review to determine if the company destroyed any documents or other evidence related to its activities in Iran.[157] Mr. Zhu provided no information that could allow the Committee to evaluate the extent of those activities, their compliance with U.S. laws, or management's involvement in the potential destruction of documents and evidence. ZTE did not answer in specific written questions from the Committee asking why it sought to limit its

Iranian business activities; whether ZTE will honor its current contracts in Iran; or whether those contracts include training or maintenance of surveillance equipment. Further, ZTE refused to answer questions about what products ZTE resold in Iran. ZTE also refused to provide any documents on its activities in Iran.

v. ZTE failed to provide clear answers to Committee questions about its R&D activities, particularly as they relate to any military or government projects.

Given ZTE's background, the Committee was interested in ZTE's R&D activities, and particularly its R&D activities with or on behalf of the Chinese military or security services. This information would help the Committee evaluate whether a company seeking to build critical infrastructure in the United States could also be working with the Chinese government on R&D projects with the purpose of finding or exploiting vulnerabilities in those systems.

ZTE's known connections to Chinese government-related research institutes are of particular interest. For example, ZTE acknowledges that one of its primary shareholders, Zhongxingxin, is owned in part by Xi'an Microelectronics, a subsidiary of China Aerospace Electronics Technology Research Institute, a state-owned research institute.[158] Another 17% of Zhongxingxin is held by Aerospace Guangyu, a subsidiary of a state-owned enterprise whose business includes production of, among other things, aerospace technology products.[159] ZTE failed to answer questions from the Committee seeking further details about the range of products theses research institute have produced on the Chinese government so the committee could not evaluate whether those technologies were produced for military or intelligence purposes.[160]

These ties to Chinese government research institutes and production companies, the Committee sought more information on the details of ZTE's R&D activities, and particularly its potential work on behalf of the government, military, or security services. ZTE was proud to explain that it had established 18 state-of-the-art R&D centers throughout China, France, and India, and to employ over 30,000 research professionals. ZTE further claims that 10% of the company's annual revenue is invested in R&D. ZTE failed, however, to answer Committee questions about the technologies it may create or sell to the Chinese government and military. During the Committee's April 12, 2012 meeting with company officials, Mr. Steinert, the independent board member, stated that, ZTE's work on behalf of the Chinese telecommunications providers that happen to be state-owned enterprises does not suggest that ZTE does work on behalf of the military or intelligence services. When providing written answers ZTE refused to provide clear answers about the nature and extent of any work it does on behalf of the Chinese military or security services. Rather, ZTE states that "[t]he funding ZTE has received from government or consortia during the past several years is indistinguishable from similar funding available throughout the world in companies engaged in R&D through normal procurement processes."[161]

To the extent ZTE's R&D activities are simply in response to standard government procurement processes, the Committee does not understand why it refuses to answer direct questions about the details of those projects. For this reason, the Committee cannot allay concerns that ZTE is aligned with Chinese military and intelligence activities or research institutes.

CONCLUSION AND RECOMMENDATIONS

The Committee launched this investigation to seek answers to some persistent questions about the Chinese telecommunications companies Huawei and ZTE and their ties to the Chinese government. Throughout the months-long investigation, both Huawei and ZTE sought to describe, in different terms, why neither company is a threat to U.S. national-security interests. Unfortunately, neither ZTE nor Huawei have cooperated fully with the investigation, and both companies have failed to provide documents or other evidence that would substantiate their claims or lend support for their narratives.

Huawei, in particular, provided evasive, nonresponsive, or incomplete answers to questions at the heart of the security issues posed. The failure of these companies to provide responsive answers about their relationships with and support by the Chinese government provides further doubt as to their ability to abide by international rules.

RECOMMENDATIONS

Based on this investigation, the Committee provides the following recommendations:

Recommendation 1: The United States should view with suspicion the continued penetration of the U.S. telecommunications market by Chinese telecommunications companies.

- The United States Intelligence Community (IC) must remain vigilant and focused on this threat. The IC should actively seek to keep cleared private sector actors as informed of the threat as possible.
- The Committee on Foreign Investment in the United States (CFIUS) must block acquisitions, takeovers, or mergers involving Huawei and ZTE given the threat to U.S. national security interests. Legislative proposals seeking to expand CFIUS to include purchasing agreements should receive thorough consideration by relevant Congressional committees.
- U.S. government systems, particularly sensitive systems, should not include Huawei or ZTE equipment, including in component parts. Similarly, government contractors – particularly those working on contracts for sensitive U.S. programs – should exclude ZTE or Huawei equipment in their systems.

Recommendation 2: Private-sector entities in the United States are strongly encouraged to consider the long-term security risks associated with doing business with either ZTE or Huawei for equipment or services. U.S. network providers and systems developers are strongly encouraged to seek other vendors for their projects. Based on available classified and unclassified information, Huawei and ZTE cannot be trusted to be free of foreign state influence and thus pose a security threat to the United States and to our systems.

Recommendation 3: Committees of jurisdiction within the U.S. Congress and enforcement agencies within the Executive Branch should investigate the unfair trade practices of the Chinese telecommunications sector, paying particular attention to China's continued financial support for key companies.

Recommendation 4: Chinese companies should quickly become more open and transparent, including listing on western stock exchange with advanced transparency requirements, offering more consistent review by independent third-party evaluators of their financial information and cyber-security processes, complying with U.S. legal standards of information and evidentiary production, and obeying all intellectual-property laws and standards. Huawei, in particular, must become more transparent and responsive to U.S. legal obligations.

Recommendation 5: Committees of jurisdiction in the U.S. Congress should consider potential legislation to better address the risk posed by telecommunications companies with nation-state ties or otherwise not clearly trusted to build critical infrastructure. Such legislation could include increasing information sharing among private sector entities, and an expanded role for the CFIUS process to include purchasing agreements.

End Notes

[1] Ken Hu, "Huawei Open Letter." http://online.wsj.com/public/resources/documents/Huawei 20110205.pdf (accessed August 2, 2012).

[2] Huawei's letter was issued in February, 2011, when the Committee on Foreign Investment in the United States (CFIUS) issued a recommendation that Huawei voluntarily divest assets it received in a 2010 deal with 3Leaf, a United States company that developed advanced computer technologies. Shayndi Raice, "Panel Poised to Recommend Against Huawei Deal," *Wall Street Journal*, February, 11, 2011. http://www.wsj.com/article/SB20001424052748704629004576136340771329706.html (accessed August 2, 2012)

[3] A classified annex to this report provides both classified information relevant to the discussion, as well as information about the resources and priorities of the IC.

[4] Steven M. Rinaldi, James P. Peerenboom, and Terrence K. Kelly, "Identifying, Understanding, and Analyzing Critical Infrastructure Interdependencies," *IEEE Control Systems Magazine*, December 2001.

[5] "The former National Counterintelligence Executive, Mr. Robert Bryant, recently noted that, 'Insider threats remain the top counterintelligence challenge to our community.' An insider threat arises when a person with authorized access to U.S. Government resources, to include personnel, facilities, information, equipment, networks, and systems, uses that access to harm the security of the United States. Malicious insiders can inflict incalculable damage. They enable the enemy to plant boots behind our lines and can compromise our nation's most important endeavors. Over the past century, the most damaging U.S. counterintelligence failures were perpetrated by a trusted insider with ulterior motives." http://www.ncix.gov/issues/ithreat/index.php

[6] FBI, *Intelligence Bulletin*, "Supply Chain Poisoning: A Threat to the Integrity of Trusted Software and Hardware," June 27, 2011: 1.

[7] Office of National Counterintelligence Executive, *Report to Congress on Foreign Economic Collection and Industrial Espionage*, "Foreign Spies Stealing US Economic Secrets in Cyberspace."(October 2011, Washington, DC: 1.)

[8] United States Congress, *2011 Annual Report of U.S.-China Economic and Security Review*. (2011, Washington DC: 59.)

[9] National Institute of Standards and Technology, *Draft NISTIR 7622*, "Piloting Supply Chain Risk Management for Federal Information Systems," June 2010, 28.

[10] Joint Press Conference, March, 29, 2012, Sydney, Australia. http://www.pm.gov.au/press-office/transcript-joint-press-conference-sydney-1.

[11] The Economist, "Huawei: The Company that Spooked the World," Economist, August, 4, 2012. http://www.economist.com/node/21559929 (accessed September 30, 2012).

[12] United States Congress, *2011 Annual Report of U.S.-China Economic and Security Review*. (2011, Washington DC: 148.)

[13] Office of National Counterintelligence Executive, *Report to Congress on Foreign Economic Collection and Industrial Espionage*, "Foreign Spies Stealing US Economic Secrets in Cyberspace."(October 2011, Washington, DC: i.)

[14] Ibid, 5; HPSCI staff interviews with cyber-security experts.

[15] Ibid, 5.

[16] Defense Science Board, *Report on Mission Impact of Foreign Influence on DoD Software*, September 2007: viii.

[17] "Where State security requires, a State security organ may inspect the electronic communication instruments and appliances and other similar equipment and installations belonging to any organization or individual." State-Security Law of the People's Republic of China, Article 11.

[18] Defense Science Board, *Report on Mission Impact of Foreign Influence on DoD Software*, September 2007: viii.

[19] Northrop Grumman Corp, *Occupying the Information High Ground: Chinese Capabilities for Computer Network Operations and Cyber Espionage*, prepared for U.S.-China Economic and Security Review Commission, March 7, 2012, 6-8.

[20] The Economist, "The Long March of the Invisible Mr. Ren," *the Economist*, June 2, 2011. http://www.economist.com/node/18771640 (accessed on September 15, 2012).

[21] FBI, *Intelligence Bulletin*, "Supply Chain Poisoning: A Threat to the Integrity of Trusted Software and Hardware," June 27, 2011: 4.

[22] ZTE, *Submissions to HPSCI*, July 3, 2012; ZTE, *Submission to HPSCI*, August 3, 2012; Ken Hu, "Huawei Open Letter." http://online.wsj.com/public/resources/documents/Huawei 20110205.pdf (accessed August 2, 2012). John Suffolk, Huawei's Global Security Officer, previously served as the Chief Information Officer with the UK government at a time when the UK entered into its agreement with Huawei to set up the Cyber Security Evaluation Center (CSEC). Mr. Suffolk advocated for a cyber-security and supply-chain solution that would recognizing the issues as a global concern that must be addressed at an international level, preferably by an international standards-setting organization through which all products must pass. Mr. Suffolk also highlighted that in the present age, technology is moving faster than our ability to adapt our institutions. Key assumptions are that security requires a whole systems approach, and that all systems will be breached at some point. Thus, in Mr. Suffolk's view, telecommunications companies and governments must manage the risk, focus on areas of most concern, instill diversity and adaptability, and learn to deal with the consequences. Mr. Suffolk acknowledged that Huawei's desire to be an end-to-end provider for whole network solutions does not align with his proposed solutions to the supply-chain concerns, which depend on diversity of supply. HPSCI meeting with John Suffolk, February 23, 2012.

[23] Anderson, R., & Fuloria, S. Certification *and Evaluation: A Security Economics Perspective. Emerging Technologies and Factory Automation*, (2009).

[24] Ken Thompson, *Reflections on Trusting Trust. Turing Award Lecture*, (1984).

[25] Gerwin Klein, Formal Verification of an OS Kernel. *Symposium on Operating Systems Principles*. Big Sky, MT, USA: Association of Computing Machinery, (2009).

[26] Daniel Jackson, Martyn Thomas, and Lynette I. Millett, Eds. *Software for Dependable Systems: Sufficient Evidence? Committee on Certifiably Dependable Software Systems*, National Research Council. (National Academies Press, 2007.)

[27] Rules of the House of Representatives, 112th Congress, Rules 10(3)(m), 11.

[28] Understanding and developing a strategy to protect the country from Chinese cyber espionage in the United States is one of the obligations of U.S. counterintelligence professionals. Many reports have suggested that the Intelligence Community continues to struggle integrating and acting on its counterintelligence mission. As Michelle Van Cleave, former head of the National Counterintelligence Executive, has explained, "the U.S. government

has been slow to appreciate the effects of foreign intelligence operations, much less to address the threats they pose to current U.S. foreign policy objectives or enduring national security interests." Michelle Van Cleave, "Chapter 2: The NCIX and the National Counterintelligence Mission: What Has Worked, What Has Not, and Why," in *Meeting Twenty-First Century Security Challenges*, 62.

[29] Office of National Counterintelligence Executive, *Report to Congress on Foreign Economic Collection and Industrial Espionage*, "Foreign Spies Stealing US Economic Secrets in Cyberspace."(October 2011, Washington, DC)

[30] In response to the Committee's June 12, 2012, document request, ZTE provided one document: a summary of its cyber-security measures. Huawei provided no documents other than materials already on the company's website or otherwise publicly released. After the September 13, 2012 hearing, Huawei provided a document labeled "Internal Compliance Program (ICP)," dated March 2012. That document summarizes Huawei's internal policy with respect to trade control policies. Huawei provided no material that would allow the Committee to evaluate their compliance with or enforcement of that policy. Huawei also provided a copy of the publicly released paper entitled "Cyber Security Perspectives" prepared by John Suffolk, and Huawei's public statement regarding its Commercial Operations in Iran.

[31] House Permanent Select Committee on Intelligence, *Hearing on Investigation of the Security Threat Posed by Chinese Telecommunications Companies Huawei and ZTE*, 112th Congress, 2nd session (September 13, 2012).

[32] Given the sensitivities involved, and to protect these witnesses from retaliation or dismissal, the Committee decided to keep the identities of these individuals confidential.

[33] As the U.S.-China Commission has highlighted, even the largely circumstantial evidence that known incidents appear state sponsored is compelling -- as the actors' targeting often focuses on key defense and foreign-policy sources of information, which are of most concern to the state and not commercial entities.

[34] United States Congress, *2011 Annual Report of U.S.-China Economic and Security Review*. (2011, Washington DC: 59.)

[35] United States Congress, *2011 Annual Report of U.S.-China Economic and Security Review*. (2011, Washington DC: 59-60.)

[36] ZTE, *Submission to HPSCI*, July 3, 2012, 3.

[37] Discussion with PLA Piper, June 2012. Huawei, in its responses to Questions for the Record after the September 13, 2012, hearing, denied that there is any state-secret concern with their documentation. The Committee is left wondering, then, why Huawei has refused to provide internal documentation that could substantiate its claims. Moreover, Huawei's failure to provide the list of individuals on Huawei's Chinese Communist Party Committee is an example in which the Committee believes the state's concerns with state secrets is particularly relevant. Huawei's continuous failure to provide such information cannot be explained otherwise.

[38] Huawei Investment & Holding Co., Ltd., *2011 Annual Report*, 7.

[39] Ken Hu, "Huawei Open Letter." http://online.wsj.com/public/resources/documents/Huawei 20110205.pdf (accessed August 2, 2012).

[40] That report suggests that Huawei "was founded in 1988 by Ren Zhengfei, a former director of the PLA General Staff Department's Information Engineering Academy, which is responsible for telecom research for the Chinese military. Huawei maintains deep ties with the Chinese military, which serves a multi-faceted role as an important customer, as well as Huawei's political patron and research and development partner. Both the government and the military tout Huawei as a national champion, and the company is currently China's largest, fastest-growing, and most impressive telecommunications-equipment manufacturer. Evan Medeiros et al., *A New Direction for China's Defense Industry*, Rand Corporation: 218-219. http://www.rand.org/pubs/monographs/2005/RAND_MG334.pdf.

[41] Ibid, 217-219

[42] The Economist, "Huawei: The Company that Spooked the World," *Economist*, August, 4, 2012. http://www.economist.com/node/21559929 (accessed September 30, 2012).

[43] Juha Saarinen, "Analysis: Who Really Owns Huawei?," *ITNews*, May 28, 2012.

[44] Ken Hu, "Huawei Open Letter." http://online.wsj.com/public/resources/documents/Huawei 20110205.pdf (accessed August 2, 2012).

[45] Richard McGregor, *The Party: The Secret World of China's Communist Rulers*, 2010: 204.

[46] Huawei, *Submission to HPSCI*, July 3, 2012.

[47] House Permanent Select Committee on Intelligence, *Hearing on Investigation of the Security Threat Posed by Chinese Telecommunications Companies Huawei and ZTE*, 112th Congress, 2nd session (September 13, 2012).

[48] Huawei Investment & Holding Co., Ltd., *2011 Annual Report*, 2.

[49] Huawei, *September 25, 2012 Response to Questions for the Record*, at __.

[50] Huawei, *September 25, 2012 Responses to Questions for the Record*, 6-7.

[51] Interviews with Huawei officials, February 23, 2012.

[52] Interviews with Huawei officials, February 23, 2012.

[53] Interviews with Huawei officials, February 23, 2012.

[54] Mike Rogers and Dutch Ruppersburg, letter to Huawei, June 12, 2012; Huawei, *letter to HPSCI*, "Response to June 12, 2012 Letter," July 3, 2012.

[55] Huawei, *Documents Provided in Advance of February 23, 2012 entitled Shareholder Agreements*.

[56] John Lee, "The Other Side of Huawei," *Business Spectator*, March 30, 2012.

[57] United States Congress, *2011 Annual Report of U.S.-China Economic and Security Review*. (2011, Washington DC: 59)

[58] Ibid.

[59] House Permanent Select Committee on Intelligence, *Hearing on Investigation of the Security Threat Posed by Chinese Telecommunications Companies Huawei and ZTE*, 112th Congress, 2nd session (September 13, 2012).

[60] Huawei, *Submission to House Permanent Select Committee on Intelligence*, July 3, 2012, 1.

[61] Ibid.

[62] Ibid.

[63] Huawei, *July 2, 2012 Submission*, 7-15

[64] Highlighting that as China moved from a pure control economy in the 1990s, Chinese companies experienced particular difficulties raising capital in foreign capital markets, including the "most sensitive of all, how would they explain the role of the internal party bodies, which for years had run companies, free of any of the inconvenient structuring of corporate reporting and governance rules." Richard McGregor, *The Party: The Secret World of China's Communist Rulers*, 2010: 47; See John Lee, "The Other Side of Huawei," Business Spectator, March 30, 2012

[65] Huawei, *Submission to House Permanent Select Committee on Intelligence*, July 3, 2012, 2.

[66] Ibid.

[67] *See* John Lee, "The Other Side of Huawei," *Business Spectator*, March 30, 2012; Richard McGregor, *The Party: The Secret World of China's Communist Rulers*, 2010.

[68] Richard McGregor, *The Party: The Secret World of China's Communist Rulers*, 2010: 72.

[69] Meeting with Mr. Ren, May 23, 2012.

[70] Huawei, *Submission to House Permanent Select Committee on Intelligence*, July 3, 2012.

[71] Huawei officials stated that China had cancelled ranking system at the time. HPSCI Interviews with Huawei officials, February 23, 2012.

[72] Huawei officials suggested that the rumors that Mr. Ren is a former PLA General is the result of confusion with Julong, another Chinese telecommunications company and state-owned enterprise whose President is a Major General in the PLA. HPSCI Interviews with Huawei officials, February 13, 2012.

[73] Huawei, *September 25, 2012 Responses to HPSCI Questions for the Record*, 8.

[74] Huawei, *September 25, 2012 Responses to HPSCI Questions for the Record*, 8.

[75] Huawei asserted that Chen Jinyang, who invested 3,500 RMB, was a 26-year-old manager at the Chinese Trade Department.
[76] Interviews with Huawei officials, February 23, 2012.
[77] Interviews with Huawei officials, February 23, 2012.
[78] Interview with Ken Hu, February 23, 2012.
[79] Scholars of the Chinese political economy suggest that national champions are those chosen by China to be supported both financially and otherwise by the state because of the strategic importance of the sector and the company to China's national interests. See John Lee, "The Other Side of Huawei," *Business Spectator*, March 30, 2012
[80] Huawei, *Submission to House Permanent Select Committee on Intelligence*, July 3, 2012, 19.
[81] Ibid.
[82] Interviews with Huawei officials, February 23, 2012; Huawei presentation, February 23, 2012.
[83] Interviews with Huawei officials, February 23, 2012
[84] Huawei, *Submission to House Permanent Select Committee on Intelligence*, July 3, 2012, 19-20.
[85] Mike Rogers and Dutch Ruppersburg, letter to Huawei, June 12, 2012, 6.
[86] Huawei, *Submission to House Permanent Select Committee on Intelligence*, July 3, 2012, 19-20.
[87] Phone conversations with Huawei representatives, June 2012.
[88] Huawei, *Submission to House Permanent Select Committee on Intelligence*, July 3, 2012, 20-21.
[89] John Lee, "The Other Side of Huawei," *Business Spectator*, March 30, 2012.
[90] The Economist, Huawei: The Company that Spooked the World," *Economist*, August, 4, 2012. http://www.economist.com/node/21559929 (accessed September 30, 2012);
[91] House Permanent Select Committee on Intelligence, *Hearing on Investigation of the Security Threat Posed by Chinese Telecommunications Companies Huawei and ZTE*, 112th Congress, 2nd session (September 13, 2012).
[92] Huawei, *Slide Presentation dated November 2011*, 8.
[93] Huawei, *Responses to HPSCI Questions for the Record*, September 25, 2012, 1.
[94] Ibid.
[95] Ibid.
[96] Inteviews with Huawei officials, February 23, 2012
[97] Huawei, *Corporate Presentation*, February 23, 2012, 26.
[98] Huawei, *Submission to House Permanent Select Committee on Intelligence*, July 3, 2012, 2.
[99] Huawei, *Corporate Presentation*, February 23, 2012, 27.
[100] Huawei, *Responses to HPSCI Questions for the Record*, September 25, 2012, 2.
[101] Inteviews with Huawei officials, February 23, 2012.
[102] House Permanent Select Committee on Intelligence, *Hearing on Investigation of the Security Threat Posed by Chinese Telecommunications Companies Huawei and ZTE*, 112th Congress, 2nd session (September 13, 2012).
[103] Huawei, *Submission to House Permanent Select Committee on Intelligence*, July 3, 2012, 21.
[104] Interviews with Huawei officials, February 23, 2012.
[105] Ren Zhengfei, speech at Huawei BT Division & Huawei UK, June 30, 2007, quoted in Huawei magazine *Improvement*, Issue 58.
[106] The Commerce Department, working with the Defense Department, has sought information from the private sector to better understand the entire scope of cyber-risks facing the country's critical telecommunication infrastructure. The Commerce issued a survey under the Defense Production Act to dozens of U.S. based companies to gather better information on the security of their networks. The review of that information is still ongoing.
[107] The Committee has offered on numerous occasions to provide Huawei an opportunity to provide the information the Committee needs to evaluate the security of U.S. networks in a closed forum or under an agreement to provide such information confidentially. Huawei has continuously refused to accept any such offer, option instead to simply assert that such

details are confidential. The Committee intends to continue evaluating these issues and plans to approach Huawei in the future for more details on these contracts to fulfill the Committee's duty to evaluate the risk posed by these firms.

[108] House Committee on Foreign Affairs, *Hearing on Unfair Trade Practices against the US*, 112th Congress, 2nd session (July 19, 2012).

[109] Interview with Huawei officials, February 23, 2012.

[110] Interview with Employees.

[111] John Lee, "The Other Side of Huawei," *Business Spectator*, March 30, 2012.

[112] Interview with Huawei officials, February 23, 2012.

[113] Huawei, *Submission to House Permanent Select Committee on Intelligence*, July 3, 2012.

[114] Interview with Huawei Employees.

[115] Interview with Huawei Employees.

[116] Interview with industry experts.

[117] Huawei representatives admitted to Committee staff that using this presentation was in violation of McKinsey's copyright protections, and that McKinsey and Huawei have no business relationship thus undermining any claim that Huawei had a right to use the slide. Huawei, *Slide Presentation dated November 2011*, 8 (using McKinsey & Co. material).

[118] Interview with Huawei Officials, February 13, 2012.

[119] Ibid.

[120] Marguerite Reardon, "Huawei Admits Copying," *Light Reading*, March 25, 2003. http://www.lightreading.com/document.asp?doc_id=30269 (accessed on August 13, 2012)

[121] Ibid.

[122] House Permanent Select Committee on Intelligence, *Hearing on Investigation of the Security Threat Posed by Chinese Telecommunications Companies Huawei and ZTE*, 112th Congress, 2nd session (September 13, 2012).

[123] Huawei, *Submission to House Permanent Select Committee on Intelligence*, July 3, 2012, 6.

[124] Ibid.

[125] Ibid, 5-6.

[126] Ibid, 3-4.

[127] Ibid, 3.

[128] Interviews with Huawei officials, February 23, 2012.

[129] Huawei, *September 25, 2012 Responses to Questions for the Record*, 12.

[130] Ibid.

[131] Internal Huawei email, dated July 1, 2011.

[132] Ibid.

[133] Interviews with former Huawei employees.

[134] Interviews with former Huawei employees.

[135] Huaw ei, *Slide Presentation dated November 2011*, 8.

[136] ZTE August 3, 2012 submission, at 12-17.

[137] ZTE, *Submissions to HPSCI*, August 3, 2012, 23.

[138] ZTE, *2011 Annual Report*, 68-69.

[139] "The national '12th Five Year Plan' has provided driving force for the further development of the domestic telecommunications industry." ZTE, *2011 Annual Report*, 69.

[140] Meeting with ZTE officials, April 12, 2012, Shenzhen, China.

[141] ZTE, *Submissions to HPSCI*, July 3, 2012.

[142] Ibid, 4.

[143] Ibid.

[144] As a report commissioned by the U.S. China-Commission stated: "The IT sector in China can be considered a hybrid defense industry, able to operate with success in commercial markets while meeting the demands of its military customers. The Chinese telecommunications market is heavily influenced by its largest domestic members—such as hardware and networking giants Huawei Shenzhen Technology Company, Zhongxing Telecom (ZTE), and Datang Telecom Technology Co., Limited. These companies and some smaller players

are not always directly linked to the PLA or C4ISR modernization because of their strong domestic and international commercial orientation. The digital triangle model, however, allows them to benefit directly from a background network of state research institutes and government funding in programs that do have affiliation or sponsorship of the PLA." Northrop Grumman Corp, *Occupying the Information High Ground: Chinese Capabilities for Computer Network Operations and Cyber Espionage*, prepared for U.S.-China Economic and Security Review Commission, March 7, 2012, 69.

[145] ZTE, *Submissions to HPSCI*, July 3, 2012, 2.

[146] Ibid, 9

[147] Ibid.

[148] Ibid, 4.

[149] House Permanent Select Committee on Intelligence, *Hearing on Investigation of the Security Threat Posed by Chinese Telecommunications Companies Huawei and ZTE*, 112th Congress, 2nd session(September 13, 2012).

[150] John Merrigan, letter to Katie Wheelbarger, September 28, 2012.

[151] Affidavit of Timothy Steinert, at para. 6.

[152] ZTE, *Submissions to HPSCI*, August 3, 2012, 5.

[153] Meeting with ZTE officials, April 12, 2012, Shenzen, China.

[154] Ibid.

[155] House Permanent Select Committee on Intelligence, *Hearing on Investigation of the Security Threat Posed by Chinese Telecommunications Companies Huawei and ZTE*, 112th Congress, 2nd session (September 13, 2012).

[156] Ellen Nakashima, "Chinese telecom firm ZTE probed for alleged sale of U.S. surveillance equipment to Iran," *Washington Post*, July 13, 2012. http://www.washingtonpost.com/world/national-security/chinese-telecom-firm-zte-probed-for-alleged-sale-of-us-surveillance-equipment-to-iran/2012/07/13/gJQA6mKUiW_story.html.

[157] House Permanent Select Committee on Intelligence, *Hearing on Investigation of the Security Threat Posed by Chinese Telecommunications Companies Huawei and ZTE*, 112th Congress, 2nd session (September 13, 2012).

[158] ZTE, *Submissions to HPSCI*, April 2012, 4.

[159] Ibid.

[160] Ibid.

[161] ZTE, *Submissions to HPSCI*, July 3, 2012, 17.

In: Security Issues ...
Editor: Simon Johnston
ISBN: 978-1-62417-423-0

Chapter 2

STATEMENT OF CHAIRMAN MIKE ROGERS, HOUSE PERMANENT SELECT COMMITTEE ON INTELLIGENCE. HEARING ON "INVESTIGATION OF THE SECURITY THREAT POSED BY CHINESE TELECOMMUNICATIONS COMPANIES HUAWEI AND ZTE"*

- We invited these companies today as part of our ongoing investigation into the threat posed to the United States by telecommunications equipment manufactured by companies with believed ties to the Chinese government.
- Huawei and ZTE have become dominant global players in the telecommunications market leaving the world increasingly dependent on their telecommunications goods and services.
  - They do not yet dominate the U.S. market, but they seek to expand their footprint here.
  - They reap the benefit of billions of dollars in Chinese government financing, and nicely implement Beijing's explicit desire to be dominant in what China calls a "strategic sector."
  - We have heard reports about backdoors or unexplained beaconing from the equipment sold by both companies. And our sources overseas tell us that there is a reason to question whether the

* This is an edited, reformatted and augmented version of the testimony given on September 13, 2012 before the House Permanent Select Committee on Intelligence.

companies are tied to the Chinese government or whether their equipment is as it appears.

- We have heard reports about their attempts to steal the tradesecrets of other companies, which gives them a competitive advantage and makes us question their ability to abide by any rules.

- At the same time, Chinese intelligence efforts against the United States are growing in scale, intensity, and sophistication.
  - Chinese actors are also the world's most active and persistent perpetrators of economic espionage.
  - U.S. firms and cybersecurity specialists speak about an on-going onslaught of sophisticated computer network intrusions originating in China, that are almost certainly the work of, or done at the backing of, the Chinese government.
  - Chinese intelligence services and private companies often use Chinese citizens with direct access to corporate networks to steal trade secrets and other sensitive proprietary data.
- Americans have become increasingly dependent on computer networks for every aspect of our lives. Our personal information, our banking information, our transportation infrastructure, medical records, education systems, and government all depend on computer networks.
- We must have trust and confidence in these networks. But those networks are already under attack. As General Alexander, head of U.S. Cyber Command, has recently explained, the U.S. has experienced a 17-fold increase in cyber attacks between 2009 and 2011.
  - Defending against the risk of cyber attacks becomes a bigger challenge when the system itself cannot be trusted. When the equipment and software is provided by companies we cannot trust, then we must constantly worry whether our systems our going to work against us.
  - A sophisticated nation-state actor like China has the motive to tamper with the global telecommunications supply chain, and the United States is a significant priority.
    - The ability to deny service or disrupt global systems allows a foreign regime the opportunity to exert pressure or control over critical infrastructure on which our country depends.

        - The ability to maliciously modify or steal information from government and corporate entities provides China access to expensive and time-consuming research and development that assists China's place in the world.
    - Huawei and ZTE provide a wealth of opportunities for Chinese intelligence agencies to insert malicious hardware or software implants into critical telecommunications components and systems. And under Chinese law, ZTE and Huawei would likely be required to cooperate with any request by the Chinese government to use their systems or access for malicious purposes.
    - When vulnerabilities in the equipment, such as backdoors and malicious code, can be exploited by another country, it becomes a priority national security concern.
        - Every piece of this equipment, every code of software, every update, provides that country a means to act against the United States.
        - We must trust our systems if we hope to fulfill the government's most-basic duty to maintain a defense against potential foreign aggression.
- We must get to the truth and see if these companies are tied to or influenced by the Chinese government; whether they provide a means for further economic and foreign espionage by a foreign nation-state known to be a major perpetrator of cyber espionage.
- In February 2011, Huawei issued an open letter to the U.S. government requesting a full investigation into its corporate operations to try to convince us that there is no threat.
    - We decided to give Huawei that investigation.
- In the course of the investigation, the Committee has been disappointed that the companies provided little actual evidence to ameliorate the Committee's concerns.
    - In particular, they did not provide documentation supporting or confirming their claims about their formal relationships or regulatory interaction with Chinese authorities, corporate structure, ownership, operations, or management.
    - We were willing to work with both companies, to find a reasonable way to answer our documents requests. But the companies refused, apparently because to turn over internal corporate documents would potentially violate China's state-secret laws. It is strange the internal corporate documents of

purportedly private sector firms are considered classified secrets in China. This fact alone gives us a reason to question their independence.

- We hope that this hearing finally gives us the opportunity to get fulsome answers and resolve these doubts about your companies.

In: Security Issues …
Editor: Simon Johnston
ISBN: 978-1-62417-423-0

Chapter 3

STATEMENT OF DUTCH RUPPERSBERGER, RANKING MEMBER, HOUSE PERMANENT SELECT COMMITTEE ON INTELLIGENCE. HEARING ON "INVESTIGATION OF THE SECURITY THREAT POSED BY CHINESE TELECOMMUNICATIONS COMPANIES HUAWEI AND ZTE"*

Thank you Chairman Rogers. I would like to welcome our witnesses Mr. Charles Ding from Huawei and Mr. Zhu Jinyun from ZTE. Thank you for coming today.

About a year ago, the Intelligence Committee launched an investigation into the threat posed by Chinese-owned telecommunications companies working in the United States. We launched this investigation to determine whether allowing Chinese companies increased access to our U.S. telecommunications market places us at risk for greater foreign espionage, theft of U.S. trade secrets and increased vulnerabilities to our critical infrastructure.

The fact that both companies, Huawei and ZTE, were created and headquartered in China, a country known to aggressively conducts cyber espionage, raises issues. And add to that… the fear that China, a communist

* This is an edited, reformatted and augmented version of the testimony given on September 13, 2012 before the House Permanent Select Committee on Intelligence.

country, could compel these companies to provide it information or worse yet spy on Americans using this equipment.

We are concerned about allegations that products from Huawei and ZTE are being subsidized by the Chinese government, so these companies can offer bargain basement prices to unsuspecting consumers. These consumers may have no idea about the national security implications of their purchase. We are looking at all of these issues in our investigation.

In May, I joined a Congressional delegation including Congressman Schiff, Congresswoman Bachmann, and Congressman Nunes on a trip to China where we met with the Chairman of the Board of Huawei and officials from ZTE. We felt it was important to travel overseas to meet with these officials face-to-face and get the facts we need for our investigation. We appreciate the companies meeting with us.

Chairman Rogers and I also sent very detailed letters to Huawei and ZTE requesting a lot of important information, necessary to conduct our investigation. We were disappointed with the lack of direct answers to our in-person questions and vague responses to our letter. We are here today to give them another opportunity to thoroughly and accurately answer our questions.

We already know the Chinese are aggressively:

- Hacking into our nation's networks,
- Threatening our critical infrastructure and
- Stealing secrets worth millions of dollars in intellectual property from American companies.

In fact, the United States Cyber Command estimates $300 billion worth of trade secrets are stolen every year. This jeopardizes our national security and hurts U.S. competitiveness in the world market, costing our country countless jobs and money. As you can see, this is not "political jousting" or "trade protectionism masquerading as national security," we are doing this for very valid reasons.

According to a report recently released by the National Counterintelligence Executive, 6 of the 7 cases prosecuted in 2010 under the Economic Espionage Act involved a link to China. No matter how you do the math, the economic damage from the theft of intellectual property is unacceptable. In addition, the possibility that under current Chinese law even privately owned companies can be compelled to provide information to the Chinese government is even more concerning.

On this issue, other nations have already taken action. Australia recently excluded Huawei from competing on certain contracts dealing with their National Broadband Network. The United Kingdom has implemented costly security procedures for their Huawei equipment. In addition, the European Union is considering additional investigations into Chinese telecommunications companies and considering possible restrictions.

In my role as a leader of the Intelligence Committee, I believe we also have an obligation to protect the national security framework of the United States. That is why we are holding this hearing today and conducting this investigation.

I thank our witnesses from Huawei and ZTE for coming here today and I hope they will answer our questions directly. We look forward to a thoughtful and cooperative dialogue so we can get a better understanding of how these companies operate.

Thank you. Mr. Chairman, I yield back.

In: Security Issues …
Editor: Simon Johnston
ISBN: 978-1-62417-423-0

Chapter 4

STATEMENT OF CHARLES DING, CORPORATE SENIOR VICE PRESIDENT, HUAWEI. HEARING ON "INVESTIGATION OF THE SECURITY THREAT POSED BY CHINESE TELECOMMUNICATIONS COMPANIES HUAWEI AND ZTE"*

Chairman Rogers, Ranking Member Ruppersberger, and Members of the Committee, my name is Charles Ding. I am a Corporate Senior Vice President of Huawei. We at Huawei hope that its participation on this panel is helpful as you pursue your important mission of addressing issues relating to cyber security.

HUAWEI'S GROWTH INTO A GLOBAL COMPANY

I have had the privilege to watch Huawei grow from a small start up to a large, multinational organization. We have enjoyed enormous growth for the past 25 years. While we value innovation and entrepreneurship, most of all, we value integrity.

Huawei is an independent, employee-owned company that operates in more than 140 countries. About 70% of our revenue comes from overseas

* This is an edited, reformatted and augmented version of testimony given on September 13, 2012 before the House Permanent Select Committee on Intelligence.

markets. Thanks to our employees' unflagging efforts and our customer-centric strategy, Huawei has become the second largest provider of carrier network infrastructure in the world and was ranked 351 in the Fortune Global 500 list in 2011. We are proud that Huawei was recognized by Business Week as the most influential company in 2008 and that we won the Economist's Innovation Award in 2010. Huawei has established business relationships with the majority of the top 50 telecom carriers around the world, including Vodafone, British Telecom, France Telecom, Deutsche Telekom, and Telefónica.

In the beginning, as a start-up company, we focused on an underserved market -- rural areas in China. We were the first equipment vendor to establish a sales and service network that covered over 300 regional networks. By 1998, we were able to expand into urban markets in China. By the end of 2000, our revenue had reached $1.9 billion, and we employed 17,320 people.

We experienced our most substantial growth between 2000 and 2012. We were able to expand internationally and significantly grow our business. We won our first major European contract in 2005. Our revenue grew from $1.9 billion in 2000 to $32.4 billion in 2011. And we have expanded from primarily wire line products into mobile, software, core network, and device product lines.

Because of the hard work of our employees and our customer-centric strategy, Huawei has achieved significant success. Moreover, our success is attributable to four key factors.

1. The explosive growth in the global telecommunications market in China and globally. For example, from 1990 to 2000, fixed line subscribers grew 15 times and mobile subscribers grew 500 times in China.
2. We have invested significant resources in R&D and have a large R&D work force in China and India. Huawei employs 65,000 dedicated R&D engineers, the most in the industry.

Today, Huawei is the largest patent holder in China. Over the past decade, Huawei has invested more than $15 billion in R&D and has filed more than 50,000 patent applications. Huawei also pays for the use of the intellectual property of others and has paid more than $1.2 billion in royalty fees.

3. Since 1997, we have relied on advice received from Western consulting firms like IBM and Accenture to reengineer all of our

business processes. In fact, in the last five years we have invested over $400 million in management transformation.

4. Our shareholding structure allows us to attract and retain talented employees. In 1997, Huawei established an Employee Stock Ownership Program, or ESOP. We grant stocks to high-performing employees. Each year, Huawei determines the number of shares an employee may obtain, based on that employee's job level and contribution to the company. As of December 2011, we had 65,596 shareholding employees.

The ESOP encourages entrepreneurship and hard work-- the backbone of Huawei's success. The ESOP also has allowed Huawei to obtain a substantial amount of capital funding.

Huawei is entirely owned by its employees. No third party -- including government institutions -- has any ownership interest in Huawei. Huawei's success has not been based on favoritism or subsidization by the Chinese government.

Each shareholding employee who is currently employed at Huawei is entitled both to elect and to be elected as a shareholder representative. Such shareholding employees elect 51 people to serve as representatives of the ESOP. Those 51 representatives in turn elect thirteen people to serve as members of the Board of Directors and five people to serve as members of the Supervisory Board. The current 51 ESOP representatives were elected in December 2010. Of those 51 representatives, 10 joined Huawei right after they graduated. Two of those representatives (Ren Zhengfei and Wang Kexiang) have served in the military. None of the 51 representatives has held a position at a government agency. The current members of the Board of Directors and the Supervisory Board were elected by the new 51 ESOP representatives in January 2011. Huawei provided background information as to each current member of the Board of Directors and Supervisory Board in its answers to the June 12 letter request. These factors have allowed Huawei to become the $32 billion company it is today.

CYBER SECURITY IS VITALLY IMPORTANT

Huawei recognizes the vital importance of cyber security, which is a complex, global issue. The focus should not be confined only to companies head-quartered in China. Other companies -- Alcatel-Lucent, Nokia Siemens

Networks, Ericsson, and Cisco-- have substantial operations in China, and much equipment used in U.S. networks is developed and manufactured in China.

Since cyber security is a global issue that the whole industry has to face, governments and the whole industry should work together to improve cyber security.

Ensuring cyber security is essential for Huawei's customers. It is also good for our business. Huawei intends to continue as a leading world-wide supplier of telecommunications equipment and services. As a global company that earns a large part of its revenue from markets outside of China, we know that any improper behavior would blemish our reputation, would have an adverse effect in the global market, and ultimately would strike a fatal blow to the company's business operations. Our customers throughout the world trust Huawei. We will never do anything that undermines that trust. It would be immensely foolish for Huawei to risk involvement in national security or economic espionage.

Let me be clear -- Huawei has not and will not jeopardize our global commercial success nor the integrity of our customers' networks for any third party, government or otherwise. Ever.

Huawei has taken substantial steps to improve cyber security. As a crucial company strategy, it has created and implemented a respected end to end global cyber security assurance system. It has hired as its Global Cyber Security Officer John Suffolk, the former Chief Information Officer of the United Kingdom (U.K.) government, who is in charge of Huawei's cyber security strategy and manages and supervises its implementation.

In addressing the requirements of cyber security, we have built best practices into all of our policies and processes. In this way, cyber security is not an afterthought. Instead, it is the way we conduct our daily business - it is part of our DNA.

At Huawei, we adopt the "many eyes" and "many hands" approach to ensure transparency and security. We positively encourage audits, reviews and inspections of all technology vendors, including Huawei, in a fair and non-discriminatory manner. Each audit and review enables each company to challenge its thinking, policies, and procedures, which in turn will enhance its capability, product quality, and product security. At Huawei, we have already gone the extra mile in this respect.

Huawei proposes that governments and industry leaders come together to develop a global framework for cyber security assurance. Huawei would welcome the opportunity to develop a program in coordination with the U.S.

government to address cyber security here. Huawei Is Independent Like any corporation, Huawei complies in good faith with the laws, rules, and regulations of the governments in the countries where it does business. Thus, Huawei complies with the laws, rules, and regulations of the Chinese government. However, the Chinese government has no influence over Huawei's daily operations, investment decisions, profit distributions, or staffing. Nor does the PLA. As previously stated, neither the Chinese government nor the PLA has any ownership interest in Huawei.

We manufacture only products consistent with civilian communications standards. Huawei does not engage in customized R&D or production for military purposes. And as already observed, neither the Chinese government nor the PLA is involved in Huawei's business decisions.

As discussed above, Huawei raises significant funds through its ESOP, such that Huawei is primarily self-funded. That said, Huawei supplements its resources with commercial bank loans, subject to market driven loan conditions. All of the terms and conditions of Huawei's bank loans are based on business negotiation and follow market practice, in compliance with Chinese and international financial regulations. Huawei has financing agreements with 10 Chinese banks and 23 international banks. Huawei is a strategic customer of many international banks, including HSBC, Citi, and SCB.

Huawei has an excellent credit record, without any loan default or waiver history. KPMG has been Huawei's independent auditor since 2000. Huawei's financial liabilities are a small fraction of its total assets and have little effect on its business operations.

Customer financing is a common practice across the telecommunications industry. Customers may contact Huawei seeking help in locating financing for their purchases of Huawei's products; alternatively, financial institutions may contact Huawei seeking to provide credit support for procurement by Huawei's customers. We may then either refer the customer to a financial institution, or refer the financial institution to a customer. The financial institutions will evaluate project risks in accordance with their internal credit policies and approval procedures and make independent decisions whether to extend loans to the customers and on what terms. To the best of Huawei's knowledge, the financial institutions determine the financing plan in accordance with standard loan processes. The financial institutions also independently and directly negotiate the financing agreement terms with the customers.

Huawei in the United States

Huawei has been in the U.S. for 10 years. We have 1700 employees here. Over the last decade, Huawei has been committed to bringing our innovative products and services to U.S. customers, and to enabling U.S. citizens to have the same access to our high-quality communications services as do more than 3 billion people in other parts of the world. For example, we have introduced our world-leading distributed wireless products to the U.S., which can significantly improve network performance and reduce energy consumption. We are fully aware that, in a high-end market like the U.S., we can only win customers' trust and sustain our development by providing high-quality products and services.

We are striving to become a model investor, employer, tax payer, and corporate citizen in the U.S. In 2011 alone, Huawei procured $6.6 billion of goods and services from US companies. Huawei has invested approximately $500 million in the U.S. in the last five years.

We are committed to openness, transparency, and cooperation so that people get to know Huawei better. We strive to establish good relationships based on mutual trust with customers, governments at all levels, and other sectors of society. Our goal is to create value for customers and for the broader communities in which we operate.

Although we want to expand our presence in the U.S. and thus create more employment and growth opportunities, we have been hindered by unsubstantiated, non-specific concerns that Huawei poses a security threat. We have respectfully asked HPSCI to provide us with any allegations it has that Huawei has engaged in national security or economic espionage. HPSCI has not responded to this request. If HPSCI has specific allegations that Huawei has engaged in cyber espionage, we ask it again to tell us, so that we will have the opportunity to respond and specifically address any concerns the Committee may have.

Huawei's Response to HPSCI's Requests for Information

We recognize that the Committee has been critical of Huawei's answers to the June 12 letter request. We have two main responses to that criticism.

1. Prior to the June 12 letter request, Huawei provided the Committee with a wealth of information, including very detailed written information about the company. We also have spoken with the Committee at length, in three different sessions. We welcomed the Committee into Huawei's facilities in China and made our top executives, including our president and CEO, available to talk with it.
2. Huawei did not fully respond to the June 12, 2012 letter request for several reasons. It was literally impossible, particularly in the three week time frame provided for response, for Huawei to respond completely to the sweeping requests in that letter. For instance, demands for "all documents" regarding numerous subjects from Huawei's worldwide locations were logistically impossible to fulfill. These requests sought hundreds of thousands of documents and required that each document not in English be translated. To respond fully would have been exceedingly burdensome. The requests also sought highly sensitive, proprietary business information, which, we respectfully submit, no responsible company, foreign or domestic, would voluntarily produce.

Huawei truly seeks to cooperate with the Committee. That is why I am here voluntarily to answer your questions. Indeed, I believe that the witnesses here today are the first representatives of major Chinese corporations ever to testify before a congressional committee in the U.S. But Huawei must cooperate in a way that protects its legitimate interests and those of its business partners, just as a similarly situated U.S. based company would protect such interests.

Huawei is committed to the U.S. market, the single largest telecom market in the world. Over the past decade, we diligently invested in the U.S. In the next decade, we hope that Huawei can integrate into the U.S. market. We hope that the U.S. information and communications technology industry will prosper with Huawei's participation. We also hope that the American people can enjoy faster network speed and better user experience by using our world leading technologies. We look forward to a productive discussion about the global issue of cyber security today and in the future.

In: Security Issues …
Editor: Simon Johnston
ISBN: 978-1-62417-423-0

Chapter 5

Statement of Zhu Jinyun, Senior Vice President for Europe and North America, ZTE. Hearing on "Investigation of the Security Threat Posed by Chinese Telecommunications Companies Huawei and ZTE"*

Mr. Chairman, Ranking Member Ruppersberger and Honorable Committee Members,

My name is Zhu Jinyun, ZTE's Senior Vice President for North America and Europe. I have a supporting written statement that I urge be included in the record along with my full statement.

I am pleased to appear today to represent ZTE, one of the world's remarkable companies, operating in 140 countries before one of the most important committees in the United States Congress. When I started my career as a young engineer in a corner of China called Shenzhen, I never dreamed I would have an honor like this one.

I had the pleasure of hosting a meeting for your staff in Shenzhen on April 12 and to meet with several of you for a briefing in Hong Kong on May 23. I hope someday you will have the opportunity to visit Shenzhen, and see what a miracle it has become.

---

* This is an edited, reformatted and augmented version of testimony given on September 13, 2012 before the House Permanent Select Committee on Intelligence.

To understand ZTE, it is useful to understand Shenzhen. Thirty years ago Shenzhen was a fishing village of about 300,000 people. Today Shenzhen is China's high tech center, much like Silicon Valley, with 14 million residents, most under 30 years old. When you visit Shenzhen, you will feel optimism and hope, freedom and creativity. It is in the air.

The spirit of Shenzhen was born when China's government chose Shenzhen to be the first of four Special Economic Zones (SEZ's) in 1979. The Shenzhen SEZ was created to be an experimental center for free market companies. Today, Shenzhen is home to many of China's most successful high tech companies, including ours; and the Shenzhen Stock Exchange has over 17 million registered investors. When one recalls that, in 1979, everyone worked for the state or an SOE and there were no private companies, Shenzhen's success is impressive.

The concept of the Shenzhen SEZ seems obvious today, but it was revolutionary in 1979. Allow risk takers to develop companies on the free market model. Find out if they could stand on their own and serve people better. Learn whether Chinese companies, for the first time, could change the way China does business.

ZTE's founder and Chairman, Mr. Hou Weigui, was a young engineer at the time, working in China's aerospace sector on general purpose integrated circuit work. He and six fellow engineers saw the Shenzhen SEZ's unprecedented potential and they leapt at the chance to join in this bold experiment. They left their secure positions and embarked on this new enterprise. They were pioneers. They had no stable income, but they had a vision.

That vision was that they wanted to provide universal telecommunications service to China's undeveloped and rural areas.

This was Mr. Hou's commercial vision, not something imposed or mandated by China's government. And, from the start, he and his team saw it as a private commercial enterprise. In 1985, they launched their company with an investment of 2 million RMB (about $300,000) from three investors, notably a foreign investor from Hong Kong, and set about designing their switch. They supported their research with revenue from contract work for a trading company in Hong Kong.

For the next ten years, we had success in China's rural market. In 1987, we launched our first switch, the ZX60, developed many more over the years and won national respect for our product quality. Mr. Hou's vision was broader and in 1996, he and his team decided to pursue diversified products, serve urban and international markets. This growth plan required capital, and, in 1997 ZTE became the first Chinese telecom company to list on the Shenzhen Stock Exchange.

I grew up in rural China, and like many of my colleagues, attended college, and I joined ZTE in 1998. We wanted to be part of something new and innovative. My first assignment with ZTE was to help develop our next technology – CDMA wireless systems. It was a great time for a young engineer to join ZTE. I knew ZTE's business model was focused on engineering, R&D, innovation. The company was making unprecedented moves, including becoming the first Chinese company to enter into a licensing partnership with Qualcomm – and ZTE remains one of Qualcomm's leading partners to this day. And ZTE was expanding into diversified advanced technologies including development of a mobile communications system for China and around the world.

My first project team was six people. By 2000, we had 900 people working on CDMA wireless systems. We wanted to compete in the China market which remained dominated by giant Western vendors. Even 15 years after our founding, it was still necessary for ZTE to show China's network carriers that our solutions could be trusted. We were able to serve only about 7% of China Unicom's wireless customers, with the balance distributed among Western vendors. But, as a consequence, we became the first Chinese company manufacturing CDMA wireless systems.

Once again, ZTE won a reputation for the quality of its products in the China market. Starting with a simple circuit switch, we had built a core network and applied it to CDMA. That was state of the art product development. We engineers were quite happy that our internal designs actually worked. Our commitment to innovation and product development had paid off. When the world telecom equipment market suffered a downturn in the early 2000's, ZTE's CDMA wireless systems were in demand and helped carry us through.

Of course, Mr. Hou's vision was global. In 2004, ZTE became the first Chinese company listed both on the Shenzhen and Hong Kong Stock Exchanges.

I was assigned to help develop international markets for our CDMA RAN products. We already had learned to compete in China's developing market.

Partnering with Qualcomm, we logically identified India, Africa, and other developing markets for our equipment. Over time, ZTE's wireless equipment platforms have integrated every technology from GSM, to CDMA, to WiMAX, to 3G, LTE and beyond. Our 3G capability carried us through the 2009 economic crisis. Innovation and quality have served ZTE well.

ZTE is a free market success story built on innovation: to supply unserved markets and to help pioneer next generation technologies. And, relevant to your inquiry, ZTE's path has never been government-directed. ZTE started as, and remains today, a company of telecom equipment engineers who pursue commercial opportunity and social responsibility in product innovation. In fact, importantly, *Fortune* magazine ranks ZTE number two among China companies for global corporate social responsibility in 2012.

ZTE started out as a private company with some SOE investment. Today ZTE is owned by over 140,000 public shareholders, including many of the world's leading institutional investors. ZTE is regulated on two stock exchanges, and SOE investment has been reduced to 15.68%. ZTE's Board and management are fully devoted to serving the interests of our 140,000 public shareholders. And, because most of ZTE's business is mostly international, ZTE must comply with laws in countries throughout the world.

While ZTE appreciates its position in China's expanding telecom market, ZTE is focused on its success as a multinational company. ZTE is not an SOE or government controlled. Indeed, ZTE is China's most independent, transparent, globally focused, publicly traded telecom company.

Being a global publicly traded company naturally imposes a set of broadly recognized responsibilities and business norms. ZTE's cooperation with this Committee's investigation is one example. In our view, ZTE has set a new precedent for cooperation by any Chinese company with a US congressional investigation.

Another important example is the need for multinational telecom equipment suppliers to satisfy recognized equipment standards everywhere they do business. ZTE was the first Chinese company ever to be fully certified under the most important international and US equipment standards, including ISO 27001 and NIST FIPS 140-2. Not only is ZTE and its equipment fully certified by the leading standards-setting organizations, ZTE actively participates in these organizations, helping to lead the way in advancing standards-setting throughout the world, including cyber protection standards.

I am aware of testimony, presented to Congress earlier this year, in which the leading cyber protection experts in the US Defense Department and Department of Homeland Security have advised Congress that the most effective cyber protection is universal application of these equipment standards and Trusted Delivery Models. Not only is ZTE's equipment certified according to the most advanced standards, ZTE has offered a state of the art Trusted Delivery Model to US telecom equipment purchasers since 2010. Today, ZTE has in place a Trusted Delivery Agreement with a highly regarded independent US security assessment laboratory, Electronic Warfare Associates (EWA), which provides assurance to any US carrier purchasing ZTE telecom infrastructure equipment that the equipment will be tested continuously throughout its life cycle.

The reason experts rely upon Trusted Delivery Systems for cyber protection is that, with a reliable Trusted Delivery System, the equipment can be trusted no matter who the supplier is. Trusted Delivery Systems are vendor-neutral. Also, Trusted Delivery Systems have the same effect as highway speed cameras: they deter illegal activity. ZTE is in the vanguard of supply chain risk management and Trusted Delivery.

Given all that ZTE is doing to promote cyber security, the Committee's inquiry whether ZTE may pose a threat to critical US telecom infrastructure is very disturbing for us, as you must expect. The Committee's central question has been: would ZTE grant China's government access to ZTE telecom infrastructure equipment for a cyber attack?

Mr. Chairman, let me answer emphatically: no! China's government has never made such a request. We expect the Chinese government never to make such a request of ZTE. If such a request were made, ZTE would be bound by US law.

ZTE is committed to helping this Committee and its partners in Government and industry to promote cyber security through our cooperation here and in the future. Let me raise a concern about some legislative proposals that have been proposed recently.

Your Committee has suggested there are risks in US network carriers' purchases of telecom operating equipment from foreign vendors, particularly vendors in China. As the Committee undoubtedly understands, virtually all of the telecom equipment now sold in the United States and throughout the world contains components made, in whole or in part, in China. That includes the equipment manufactured and sold by every Western vendor, much of which is made by Chinese joint venture partners and suppliers.

We respectfully suggest that the Committee's focus on ZTE, to the exclusion of the Western telecom vendors, addresses the overall issue of risk so narrowly that it omits from the Committee's inquiry the suppliers of the vast majority of equipment used in the US market. ZTE is a relatively small US telecom equipment supplier in comparison with most of the Western vendors. Sales of ZTE's telecom infrastructure equipment in the United States comprised less than $30 million in revenue last year. Two Western vendors, alone, last year provided the US market with $14 billion dollars' worth of equipment.

ZTE should not be a focus of this investigation to the exclusion of the much larger Western vendors.

We urge the Committee to consider a serious concern with several pending legislative proposals designed to exclude Chinese equipment suppliers, directly or indirectly, from the US market. One proposal would exclude the companies named in this inquiry explicitly. Others would exclude any Chinese suppliers with "ties" to China's government Respectfully, neither proposal would protect US national security as comprehensively as implementation of the Trusted Delivery System ZTE offers US carriers. Proposals that specific Chinese companies be excluded from the US market, either directly or indirectly, would constitute obvious unfair trade practices and are so narrow that they would provide no meaningful solution in support of US cyber security.

Proposals based on China government "ties" suffer from the opposite problem. While the word "ties" is undefined, it presumably is meant to be applied broadly. If so, it is readily foreseeable that every supplier of US telecom infrastructure equipment – including the Western vendors and their Chinese manufacturing partners – would be found to have "ties" to China's government. If every vendor with "ties" to China's market is excluded from the US market, where will US carriers purchase telecom infrastructure equipment?

US chip set suppliers have advocated that Congress not adopt these exclusionary measures because they would disrupt supply relationships with Chinese equipment purchasers that are vital to the US economy. ZTE alone has purchased over $14 billion in US equipment in recent years, indirectly creating over 20,000 US jobs. We expect to make even larger purchases going forward. Market exclusion proposals are counter-productive, particularly when reliable, more effective solutions are readily available.

Responsible federal agencies and US telecom carriers have increasingly come to recognize that Trusted Delivery Systems render telecom equipment trustworthy regardless of who the supplier is. Respected international commentators have come to the same conclusion.

For example, on August 4, *The Economist* published an editorial making the following points.

- Just about everybody now makes telecom equipment in China;
- Banning Chinese telecom suppliers is no guarantee of national security;

- The answer is to require greater scrutiny of everyone, not just the Chinese firms;
- Governments should ensure telecom equipment is secure no matter who makes it; and
- This will protect networks; banning companies will not.

The Economist supports the Trusted Delivery Model, and rejects market exclusion, to protect national security.

ZTE's commitment is clear. ZTE is China's most independent, transparent, regulated, globally focused, publicly traded telecom company. We are accountable under legal and social norms everywhere we do business. ZTE's ultimate success depends on our ability to serve as a trusted partner for US telecom carriers. No company has a greater stake in promoting effective US cyber security than ZTE.

And we are readily equipped to help you achieve real results quickly. Our equipment is certified and we are helping design advanced global cyber security standards. Our end-to-end Trusted Delivery System allows US carriers to use our equipment fully confident it is, and remains, safe.

Mr. Chairman, ZTE respects the Committee's responsibility to protect US national security. We came here this morning because we want to help, with the expectation that the Congress and the United States government will provide ZTE with an open and equal opportunity to compete in the United States. We believe that ZTE is the most transparent publicly owned telecom company in China, and that ZTE is the best choice to work in concert with the Congress and the United States government and to be a trusted partner with US vendors in the formation of a truly global and free market telecom community. I thank you for the opportunity to appear today, and I will be happy to answer any of your questions.

In: Security Issues …
Editor: Simon Johnston

ISBN: 978-1-62417-423-0

Chapter 6

FOREIGN SPIES STEALING US ECONOMIC SECRETS IN CYBERSPACE: REPORT TO CONGRESS ON FOREIGN COLLECTION AND INDUSTRIAL ESPIONAGE, 2009-2011*

Office of the National Couterintelligence Executive

EXECUTIVE SUMMARY

Foreign economic collection and industrial espionage against the United States represent significant and growing threats to the nation's prosperity and security.

Cyberspace—where most business activity and development of new ideas now takes place—amplifies these threats by making it possible for malicious actors, whether they are corrupted insiders or foreign intelligence services (FIS), to quickly steal and transfer massive quantities of data while remaining anonymous and hard to detect.

* This is an edited, reformatted and augmented version of an Office of the National Couterintelligence Executive publication, dated October 2011.

US Technologies and Trade Secrets at Risk in Cyberspace

Foreign collectors of sensitive economic information are able to operate in cyberspace with relatively little risk of detection by their private sector targets. The proliferation of malicious software, prevalence of cyber tool sharing, use of hackers as proxies, and routing of operations through third countries make it difficult to attribute responsibility for computer network intrusions. Cyber tools have enhanced the economic espionage threat, and the Intelligence Community (IC) judges the use of such tools is already a larger threat than more traditional espionage methods.

Economic espionage inflicts costs on companies that range from loss of unique intellectual property to outlays for remediation, but no reliable estimates of the monetary value of these costs exist. Many companies are unaware when their sensitive data is pilfered, and those that find out are often reluctant to report the loss, fearing potential damage to their reputation with investors, customers, and employees. Moreover, victims of trade secret theft use different methods to estimate their losses; some base estimates on the actual costs of developing the stolen information, while others project the loss of future revenues and profits.

Pervasive Threat from Adversaries and Partners

Sensitive US economic information and technology are targeted by the intelligence services, private sector companies, academic and research institutions, and citizens of dozens of countries.

- Chinese actors are the world's most active and persistent perpetrators of economic espionage. US private sector firms and cybersecurity specialists have reported an onslaught of computer network intrusions that have originated in China, but the IC cannot confirm who was responsible.
- Russia's intelligence services are conducting a range of activities to collect economic information and technology from US targets.
- Some US allies and partners use their broad access to US institutions to acquire sensitive US economic and technology information, primarily through aggressive elicitation and other human intelligence (HUMINT) tactics. Some of these states have advanced cyber capabilities.

Outlook

Because the United States is a leader in the development of new technologies and a central player in global financial and trade networks, foreign attempts to collect US technological and economic information will continue at a high level and will represent a growing and persistent threat to US economic security. The nature of the cyber threat will evolve with continuing technological advances in the global information environment.

- Over the next several years, the proliferation of portable devices that connect to the Internet and other networks will continue to create new opportunities for malicious actors to conduct espionage. The trend in both commercial and government organizations toward the pooling of information processing and storage will present even greater challenges to preserving the security and integrity of sensitive information.
- The US workforce will experience a cultural shift that places greater value on access to information and less emphasis on privacy or data protection. At the same time, deepening globalization of economic activities will make national boundaries less of a deterrent to economic espionage than ever.

We judge that the governments of China and Russia will remain aggressive and capable collectors of sensitive US economic information and technologies, particularly in cyberspace.

The relative threat to sensitive US economic information and technologies from a number of countries may change in response to international economic and political developments. One or more fast-growing regional powers may judge that changes in its economic and political interests merit the risk of aggressive cyber and other espionage against US technologies and economic information.

Although foreign collectors will remain interested in all aspects of US economic activity and technology, we judge that the greatest interest may be in the following areas:

- Information and communications technology (ICT), which forms the backbone of nearly every other technology.

- Business information that pertains to supplies of scarce natural resources or that provides foreign actors an edge in negotiations with US businesses or the US Government.
- Military technologies, particularly marine systems, unmanned aerial vehicles (UAVs), and other aerospace/ aeronautic technologies.
- Civilian and dual-use technologies in sectors likely to experience fast growth, such as clean energy and health care/pharmaceuticals.

Cyberspace provides relatively small-scale actors an opportunity to become players in economic espionage. Underresourced governments or corporations could build relationships with hackers to develop customized malware or remote-access exploits to steal sensitive US economic or technology information, just as certain FIS have already done.

- Similarly, political or social activists may use the tools of economic espionage against US companies, agencies, or other entities, with disgruntled insiders leaking information about corporate trade secrets or critical US technology to "hacktivist" groups like WikiLeaks.

SCOPE NOTE

This assessment is submitted in compliance with the Intelligence Authorization Act for Fiscal Year 1995, Section 809(b), Public Law 103-359, as amended, which requires that the President biennially submit to Congress updated information on the threat to US industry from foreign economic collection and industrial espionage. This report updates the *14th Annual Report to Congress on Foreign Economic Collection and Industrial Espionage, 2008* and draws primarily on data from 2009-2011.

New Focus and Additional Resources Used for This Year's Report

This report differs from previous editions in three important ways. The first and most significant is the focus. This report gives special attention to foreign collectors' exploitation of cyberspace, while not excluding other established tactics and methods used in foreign economic collection and industrial espionage. This reflects the fact that nearly all business records, research results, and other sensitive economic or technology-related information now exist primarily in digital form. Cyberspace makes it possible for foreign collectors to gather enormous quantities of information quickly and with little risk, whether via remote exploitation of victims' computer networks, downloads of data to external media devices, or e-mail messages transmitting sensitive information.

The second difference from prior reports is that, in addition to researching the large body of intelligence reporting and analysis on economic espionage produced by the Intelligence Community, the Department of Defense (DoD), and other US Government agencies, the drafters of this report consulted new sources of government information.

Third, the Office of the National Counterintelligence Executive (ONCIX) mobilized significant resources from outside the IC during the course of this study. This included outreach to the private sector and, in particular, sponsorship of a conference in November 2010 on cyber-enabled economic espionage at which 26 US Government agencies and 21 private-sector organizations were represented. ONCIX also contracted with outside experts to conduct studies of the academic literature on the cost of economic espionage and the role of the cyber "underground economy."

Definitions of Key Terms

For the purposes of this report, key terms were defined according to both legal and analytic criteria.

The *legal criteria* derive from the language in the Economic Espionage Act (EEA) of 1996 (18 USC §§ 1831-1839). The EAA is concerned in particular with economic espionage and foreign activities to acquire US *trade secrets*. In this context, trade secrets are all forms and types of financial, business, scientific, technical, economic, or engineering information, including patterns, plans, compilations, program devices, formulas, designs, prototypes, methods, techniques, processes, procedures, programs, or codes, whether tangible or intangible, and whether stored or unstored, compiled, or memorialized physically, electronically, graphically, photographically, or in writing, if the owner (the person or entity in whom or in which rightful legal or equitable title to, or license in, is reposed) has taken reasonable measures to keep such information secret and the information derives independent economic value, actual, or potential from not being generally known to, and not being readily ascertainable through, proper means by the public. Activities to acquire these secrets include the following criminal offenses:

- *Economic espionage* occurs when an actor, knowing or intending that his or her actions will benefit any foreign government, instrumentality or agent, knowingly: (1) steals, or without authorization appropriates, carries away, conceals, or obtains by deception or fraud a trade secret; (2) copies, duplicates, reproduces, destroys, uploads, downloads, or transmits that trade secret without authorization; or (3) receives a trade secret knowing that the trade secret had been stolen, appropriated, obtained or converted without authorization (Section 101 of the EEA, 18 USC § 1831).
- *Industrial espionage,* or theft of trade secrets, occurs when an actor, intending or knowing that his or her offense will injure the owner of a trade secret of a product produced for or placed in interstate or foreign commerce, acts with the intent to convert that trade secret to the economic benefit of anyone other than the owner by: (1) stealing, or without authorization appropriating, carrying away, concealing, or obtaining by deception or fraud information related to that secret; (2) copying, duplicating, reproducing, destroying, uploading, downloading, or otherwise transmiting that information without authorization; or (3) receiving that information knowing that that information had been stolen, appropriated, obtained or converted without authorization (Section 101 of the EEA, 18 USC § 1832).

The following definitions reflect the experience of IC cyber, counterintelligence, and economic analysts:

- *Cyberspace* is the interdependent network of information technology (IT) infrastructures, and includes the Internet, telecommunications networks, computer systems, and embedded processors and controllers in critical industries.
- *Sensitive* is defined as information or technology (a) that has been classified or controlled by a US Government organization or restricted in a proprietary manner by a US corporation or other institution, or (b) that has or may reasonably be expected to have military, intelligence, or other uses with implications for US national security, or (c) that may enhance the economic competitiveness of US firms in global markets.

Contributors

ONCIX compiled this report using inputs and reporting from many US Government agencies and departments, including the Air Force Office of Special Investigations (AFOSI), Army Counterintelligence Center (ACIC), Central Intelligence Agency (CIA), Defense Intelligence Agency (DIA), Defense Security Service (DSS), Department of Energy (DoE), Department of Health and Human Services (HHS), Department of State (DoS), Federal Bureau of Investigation (FBI), National Geospatial-Intelligence Agency (NGA), National Reconnaissance Office (NRO), National Security Agency (NSA), and Naval Criminal Investigative Service (NCIS).

US Technologies and Trade Secrets at Risk in Cyberspace

The pace of foreign economic collection and industrial espionage activities against major US corporations and US Government agencies is accelerating.

FIS, corporations, and private individuals increased their efforts in 2009-2011 to steal proprietary technologies, which cost millions of dollars to develop and represented tens or hundreds of millions of dollars in potential profits.

The computer networks of a broad array of US Government agencies, private companies, universities, and other institutions—all holding large volumes of sensitive economic information—were targeted by cyber espionage; much of this activity appears to have originated in China.

Increasingly, economic collection and industrial espionage occur in cyberspace, reflecting dramatic technological, economic, and social changes that have taken place in recent years in the ways that economic, scientific, and other sensitive information is created, used, and stored. Today, nearly all business records, research results, and other sensitive economic data are digitized and accessible on networks worldwide. Cyber collection can take many forms, including: simple visits to a US company's website for the collection of openly available information; a corporate insider's downloading of proprietary information onto a thumb drive at the behest of a foreign rival; or intrusions launched by FIS or other actors against the computer networks of a private company, federal agency, or an individual.

The Appeal of Collecting in Cyberspace

Cyberspace is a unique complement to the espionage environment because it provides foreign collectors with relative anonymity, facilitates the transfer of a vast amount of information, and makes it more difficult for victims and governments to assign blame by masking geographic locations.

Security and attribution. Collectors operating in a cyber environment can collect economic information with less risk of detection. This is particularly true for remote computer network exploitation (CNE). Foreign collectors take advantage of the fact that it is difficult to detect and to attribute responsibility for these operations. There is increasing similarity between the tools, tactics, and techniques used by various actors, which reduces the reliability of using these factors to identify those responsible for computer network intrusions.

- The proliferation of malicious software (malware) presents opportunities for intelligence services and other actors to launch operations with limited resources and without developing unique tools that can be associated with them.
- Hacker websites are prevalent across the Internet, and tool sharing is common, causing intrusions by unrelated actors to exhibit similar technical characteristics.
- FIS and other foreign entities have used independent hackers at times to augment their capabilities and act as proxies for intrusions, thereby providing plausible deniability.
- Many actors route operations through computers in third countries or physically operate from third countries to obscure the origin of their activity.

Another factor adding to the challenge of attribution is the diverging perspectives of the actual targets of economic espionage in cyberspace.

- At a conference sponsored by ONCIX in November 2010, US private industry representatives said they saw little difference between cybercrime—for example, identity theft or the misappropriation of intellectual property such as the counterfeiting of commercial video or audio recordings—and the collection of economic or technology information by intelligence services or other foreign entities. Private sector organizations are often less concerned with attribution and focus instead on damage control and prevention; moreover, few companies have the ability to identify cyber intruders.

- US Government law enforcement and intelligence agencies, on the other hand, seek to establish attribution as part of their mission to counter FIS and other clandestine information collectors. They, unlike companies, also have the intelligence collection authorities and capabilities needed to break multiple layers of cover and to establish attribution where possible.

Cyberspace also offers greater security to the perpetrator in cases involving insiders. Although audits or similar cyber security measures may flag illicit information downloads from a corporate network, a malicious actor can quickly and safely transfer a data set once it is copied. A physical meeting is unnecessary between the corrupted insider and the persons or organizations the information is being collected for, reducing the risk of detection.

Faster and cheaper. Cyberspace makes possible the near instantaneous transfer of enormous quantities of economic and other information. Until fairly recently, economic espionage often required that insiders pass large volumes of documents to their handlers in physical form—a lengthy process of collection, collation, transportation, and exploitation.

- Dongfan Chung was an engineer with Rockwell and Boeing who worked on the B-1 bomber, space shuttle, and other projects and was sentenced in early 2010 to 15 years in prison for economic espionage on behalf of the Chinese aviation industry. At the time of his arrest, 250,000 pages of sensitive documents were found in his house. This is suggestive of the volume of information Chung could have passed to his handlers between 1979 and 2006.[a] The logistics of handling the physical volume of these documents—which would fill nearly four 4-drawer filing cabinets— would have required considerable attention from Chung and his handlers. With current technology, all the data in the documents hidden in Chung's house would fit onto one inexpensive CD.[b]

Extra-territoriality. In addition to the problem of attribution, it often is difficult to establish the geographic location of an act of economic espionage that takes place in cyberspace. Uncertainty about the physical location of the act provides cover for the perpetrators and complicates efforts by US Government law enforcement or intelligence agencies to respond.

Non-Cyber Methods of Economic Espionage

Although this assessment focuses on the use of cyber tools and the cyber environment in foreign efforts to collect sensitive US economic information and technologies, a variety of other methods also remain in use.

Requests for Information (RFI). Foreign collectors make unsolicited direct and indirect requests for information via personal contacts, telephone, e-mail, fax, and other forms of communication and often seek classified, sensitive, or export-controlled information.

Solicitation or Marketing of Services. Foreign companies seek entrée into US firms and other targeted institutions by pursuing business relationships that provide access to sensitive or classified information, technologies, or projects.

Conferences, Conventions, and Trade Shows. These public venues offer opportunities for foreign adversaries to gain access to US information and experts in dual-use and sensitive technologies.

Official Foreign Visitors and Exploitation of Joint Research. Foreign government organizations, including intelligence services, use official visits to US Government and cleared defense contractor facilities, as well as joint research projects between foreign and US entities, to target and collect information.

Foreign Targeting of US Visitors Overseas. Whether traveling for business or personal reasons, US travelers overseas—businesspeople, US Government employees, and contractors—are routinely targeted by foreign collectors, especially if they are assessed as having access to some sensitive information. Some US allies engage in this practice, as do less friendly powers such as Russia and China. Targeting takes many forms: exploitation of electronic media and devices, surreptitious entry into hotel rooms, aggressive surveillance, and attempts to set up sexual or romantic entanglements.

Open Source Information. Foreign collectors are aware that much US economic and technological information is available in professional journals, social networking and other public websites, and the media.

Large but Uncertain Costs

Losses of sensitive economic information and technologies to foreign entities represent significant costs to US national security. The illicit transfer of technology with military applications to a hostile state such as Iran or North Korea could endanger the lives of US and allied military personnel. The collection of confidential US Government economic information—whether by a potential adversary or a current ally—could undercut US ability to develop and enact policies in areas ranging from climate change negotiations to reform of financial market regulations. The theft of trade secrets from US companies by foreign economic rivals undermines the corporate sector's ability to create jobs, generate revenues, foster innovation, and lay the economic foundation for prosperity and national security.

Data on the effects of the theft of trade secrets and other sensitive information are incomplete, however, according to an ONCIX-sponsored survey of academic literature on the costs of economic espionage.

- Many victims of economic espionage are unaware of the crime until years after loss of the information.
- Even when a company knows its sensitive information has been stolen by an insider or that its computer networks have been penetrated, it may choose not to report the event to the FBI or other law enforcement agencies. No legal requirement to report a loss of sensitive information or a remote computer intrusion exists, and announcing a security breach of this kind could tarnish a company's reputation and endanger its relationships with investors, bankers, suppliers, customers, and other stakeholders.
- A company also may not want to publicly accuse a corporate rival or foreign government of stealing its secrets from fear of offending potential customers or business partners.
- Finally, it is inherently difficult to assign an economic value to some types of information that are subject to theft. It would, for example, be nearly impossible to estimate the monetary value of talking points for a meeting between officials from a US company and foreign counterparts.

The Cost of Economic Espionage to One Company

Data exist in some specific cases on the damage that economic espionage or theft of trade secrets has inflicted on individual companies. For example, an employee of Valspar Corporation unlawfully downloaded proprietary paint formulas valued at $20 million, which he intended to take to a new job in China, according to press reports. This theft represented about one-eighth of Valspar's reported profits in 2009, the year the employee was arrested.

Even in those cases where a company recognizes it has been victimized by economic espionage and reports the incident, calculation of losses is challenging and can produce ambiguous results. Different methods can be used that yield divergent estimates, which adds to the difficulty of meaningfully comparing cases or aggregating estimated losses.

- An executive from a major industrial company told ONCIX representatives in late 2010 that his company has used historical costs—tallying salaries, supplies, utilities, and similar direct expenses—to estimate losses from cases of attempted theft of its trade secrets. This method has the advantage of using known and objective data, but it underestimates the extent of losses in many cases because it does not capture the effect of lost intellectual property on future sales and profits.
- Harm is calculated in US civil court cases involving the theft of trade secrets by measuring the "lost profits" or "reasonable royalty" that a company is unable to earn because of the theft. Although this method requires subjective assumptions about market share, profitability, and similar factors, it does offer a more complete calculation of the cost than relying strictly on historical accounting data.
- Estimates from academic literature on the losses from economic espionage range so widely as to be meaningless—from $2 billion to $400 billion or more a year—reflecting the scarcity of data and the variety of methods used to calculate losses.

A Possible Proxy Measure of the Costs of Economic Espionage to the United States

New ideas are often a company's or an agency's most valuable information and are usually of greatest interest to foreign collectors. Corporate and government spending on research and development (R&D) is one measure of the cost of developing new ideas, and hence is an indicator of the value of the information that is most vulnerable to economic espionage. R&D spending has been tracked by the National Science Foundation (NSF) since 1953. For 2008, the most recent year available, the NSF calculated that US industry, the Federal Government, universities, and other nonprofit organizations expended $398 billion on R&D, or 2.8 percent of the US Gross Domestic Product.

Pervasive Threat From Intelligence Adversaries and Partners

Many states view economic espionage as an essential tool in achieving national security and economic prosperity. Their economic espionage programs combine collection of open source information, HUMINT, signals intelligence (SIGINT), and cyber operations—to include computer network intrusions and exploitation of insider access to corporate and proprietary networks—to develop information that could give these states a competitive edge over the United States and other rivals.

- China and Russia view themselves as strategic competitors of the United States and are the most aggressive collectors of US economic information and technology.
- Other countries with closer ties to the United States have conducted CNE and other forms of intelligence collection to obtain US economic and technology data, often taking advantage of the access they enjoy as allies or partners to collect sensitive military data and information on other programs.

Recent Insider Thefts of Corporate Trade Secrets with a Link to China

David Yen Lee…chemist with Valspar Corporation…between late 2008 and early 2009 used access to internal computer network to download about 160 secret formulas for paints and coatings to his own storage media…intended to take this proprietary information to a new job with Nippon Paint in Shanghai, China…arrested March 2009…pleaded guilty to one count of theft of trade secrets; sentenced in December 2010 to 15 months in prison.

Meng Hong…DuPont Corporation research chemist...in mid-2009 downloaded proprietary information on organic light-emitting diodes (OLED) to personal e-mail account and thumb drive…intended to transfer this information to Peking University, where he had accepted a faculty position; sought Chinese Government funding to commercialize OLED research…arrested October 2009…pleaded guilty to one count of theft of trade secrets; sentenced in October 2010 to 14 months in prison.

Yu Xiang Dong (aka Mike Yu)…product engineer wjth Ford Motor Company who in December 2006 accepted a job at Ford's China branch…copied approximately 4,000 Ford documents onto an external hard drive to help obtain a job with a Chinese automotive company…arrested in October 2009…pleaded guilty to two counts of theft of trade secrets; sentenced in April 2011 to 70 months in prison.

China: Persistent Collector

Chinese leaders consider the first two decades of the 21st century to be a window of strategic opportunity for their country to focus on economic growth, independent innovation, scientific and technical advancement, and growth of the renewable energy sector.

China's intelligence services, as well as private companies and other entities, frequently seek to exploit Chinese citizens or persons with family ties to China who can use their insider access to corporate networks to steal trade secrets using removable media devices or e-mail. Of the seven cases that were adjudicated under the Economic Espionage Act— both Title 18 USC § 1831 and § 1832—in Fiscal Year 2010, six involved a link to China.

US corporations and cyber security specialists also have reported an onslaught of computer network intrusions originating from Internet Protocol (IP) addresses in China, which private sector specialists call "advanced persistent threats." Some of these reports have alleged a Chinese corporate or government sponsor of the activity, but the IC has not been able to attribute many of these private sector data breaches to a state sponsor. Attribution is especially difficult when the event occurs weeks or months before the victims request IC or law enforcement help.

- In a February 2011 study, McAfee attributed an intrusion set they labeled "Night Dragon" to an IP address located in China and indicated the intruders had exfiltrated data from the computer systems of global oil, energy, and petrochemical companies. Starting in November 2009, employees of targeted companies were subjected to social engineering, spear-phishing e-mails, and network exploitation. The goal of the intrusions was to obtain information on sensitive competitive proprietary operations and on financing of oil and gas field bids and operations.
- In January 2010, VeriSign iDefense identified the Chinese Government as the sponsor of intrusions into Google's networks. Google subsequently made accusations that its source code had been taken—a charge that Beijing continues to deny.
- Mandiant reported in 2010 that information was pilfered from the corporate networks of a US Fortune 500 manufacturing company during business negotiations in which that company was looking to acquire a Chinese firm. Mandiant's report indicated that the US manufacturing company lost sensitive data on a weekly basis and that this may have helped the Chinese firm attain a better negotiating and pricing position.
- Participants at an ONCIX conference in November 2010 from a range of US private sector industries reported that client lists, merger and

acquisition data, company information on pricing, and financial data were being extracted from company networks—especially those doing business with China.

Russia: Extensive, Sophisticated Operations

Motivated by Russia's high dependence on natural resources, the need to diversify its economy, and the belief that the global economic system is tilted toward US and other Western interests at the expense of Russia, Moscow's highly capable intelligence services are using HUMINT, cyber, and other operations to collect economic information and technology to support Russia's economic development and security.

For example, the 10 Russian Foreign Intelligence Service (SVR) "illegals" arrested in June 2010 were tasked to collect economic and technology information, highlighting the importance of these issues to Moscow.[c]

US Partners: Leveraging Access

Certain allies and other countries that enjoy broad access to US Government agencies and the private sector conduct economic espionage to acquire sensitive US information and technologies. Some of these states have advanced cyber capabilities.

Outlook

Because the United States is a leader in the development of new technologies and a central player in global financial and trade networks, foreign attempts to collect US technological and economic information will remain at high levels and continue to threaten US economic security. The nature of these attempts will be shaped by the accelerating evolution of cyberspace, policy choices made by the economic and political rivals of the United States, and broad economic and technological developments.

Russian Leaders Link Intelligence Operations and Economic Interests

The SVR "must be able to swiftly and adequately evaluate changes in the international economic situation, understand the consequences for the domestic economy and…more actively protect the economic interests of our companies abroad."

—Vladimir Putin, President, Russian Federation, October 2007

"Intelligence…aims at supporting the process of modernization of our country and creating the optimal conditions for the development of its science and technology."

—Mikhail Fradkov, Director, SVR, December 2010

Source: Russian press reports.

Near Certainties

Evolving cyber environment. Over the next three to five years, we expect that four broad factors will accelerate the rate of change in information technology and communications technology in ways that are likely to disrupt security procedures and provide new openings for collection of sensitive US economic and technology information. These were identified in studies conducted by Cisco Systems and discussed at the ONCIX conference in

November 2010. At the same time, the growing complexity and density of cyberspace will provide more cover for remote cyber intruders and make it even harder than today to establish attribution for these incidents.

The first factor is a *technological shift*. According to a Cisco Systems study, the number of devices such as smartphones and laptops in operation worldwide that can connect to the Internet and other networks is expected to increase from about 12.5 billion in 2010 to 25 billion in 2015. This will cause a proliferation in the number of operating systems and endpoints that malicious actors such as foreign intelligence services or corrupt insiders can exploit to obtain sensitive information. Meanwhile, the underlying hardware and software of information systems will become more complex.

- Marketing and revenue imperatives will continue to lead IT product vendors to release products with less than exhaustive testing, which will also create opportunities for remote exploitation.

An *economic shift* will change the way that corporations, government agencies, and other organizations share storage, computing, network, and application resources. The move to a "cloud computing" paradigm—which is much cheaper for companies than hosting computer services in- house—will mean that employees will be able to work and access data anywhere and at any time, and not just while they are at the office, laboratory, or factory. Although cloud computing offers some security advantages, such as robust backup in the event of a systems disruption, the movement of data among multiple locations will increase the opportunities for theft or manipulation by malicious actors.

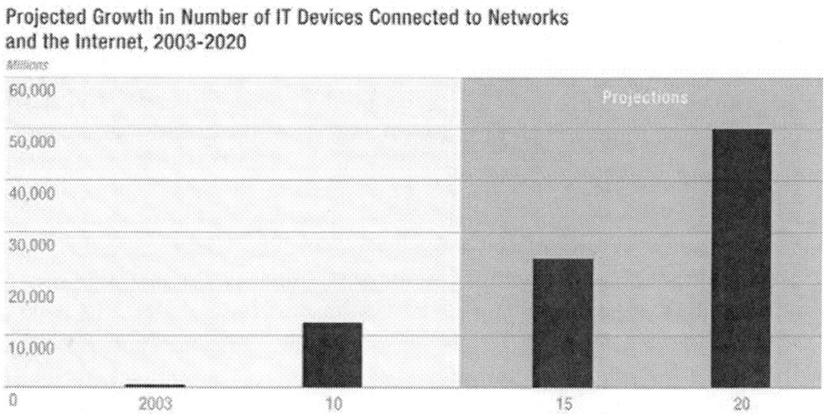

Source: CISCO Systems.

The *cultural shift* involves the rise in the US work- force of different expectations regarding work, privacy, and collaboration. Workers will tend to draw few distinctions between their home and work lives, and they will expect free access to any information they want—whether personal or professional—from any location.

- Current technology already enables many US workers to conduct business from remote locations and on-the-go at any time of day. This alteration relies on the ability of workers to connect to one another and their companies through the Internet—increasing their flexibility and corporate productivity but potentially increasing the risk of theft.

Finally, a *geopolitical shift* will continue the globalization of economic activities and knowledge creation. National boundaries will deter economic espionage less than ever as more business is conducted from wherever workers can access the Internet. The globalization of the supply chain for new—and increasingly interconnected—IT products will offer more opportunities for malicious actors to compromise the integrity and security of these devices.

Little change in principal threats. The IC anticipates that China and Russia will remain aggressive and capable collectors of sensitive US economic information and technologies, particularly in cyberspace. Both will almost certainly continue to deploy significant resources and a wide array of tactics to acquire this information from US sources, motivated by the desire to achieve economic, strategic, and military parity with the United States.

China will continue to be driven by its longstanding policy of "catching up fast and surpassing" Western powers. An emblematic program in this drive is Project 863, which provides funding and guidance for efforts to clandestinely acquire US technology and sensitive economic information. The project was launched in 1986 to enhance China's economic competitiveness and narrow the science and technology gap between China and the West in areas such as nanotechnology, computers, and biotechnology.

- The growing interrelationships between Chinese and US companies—such as the employment of Chinese-national technical experts at US facilities and the off-shoring of US production and R&D to facilities in China—will offer Chinese Government agencies and businesses increasing opportunities to collect sensitive US economic information.
- Chinese actors will continue conducting CNE against US targets.

Two trends may increase the threat from Russian collection against US economic information and technology over the next several years.

- The many Russian immigrants with advanced technical skills who work for leading US companies may be increasingly targeted for recruitment by the Russian intelligence services.
- Russia's increasing economic integration with the West is likely to lead to a greater number of Russian companies affiliated with the intelligence services—often through their employment of ostensibly retired intelligence officers—doing business in the United States.

Technologies likely to be of greatest interest. Although all aspects of US economic activity and technology are of potential interest to foreign intelligence collectors, we judge that the highest interest may be in the following areas.

Information and communications technology (ICT). ICT is a sector likely to remain one of the highest priorities of foreign collectors. The computerization of manufacturing and the push for connectedness mean that ICT forms the backbone of nearly every other technology used in both civilian and military applications.

- Beijing's Project 863, for example, lists the development of "key technologies for the construction of China's information infrastructure" as the first of four priorities.

Military technologies. We expect foreign entities will continue efforts to collect information on the full array of US military technologies in use or under development. Two areas are likely to be of particular interest:

- *Marine systems.* China's desire to jump-start development of a blue-water navy—to project power in the Taiwan Strait and defend maritime trade routes—will drive efforts to obtain sensitive US marine systems technologies.
- *Aerospace/aeronautics.* The air supremacy demonstrated by US military operations in recent decades will remain a driver of foreign efforts to collect US aerospace and aeronautics technologies. The greatest interest may be in UAVs because of their recent successful use for both intelligence gathering and kinetic operations in Afghanistan, Iraq, and elsewhere.

Civilian and dual-use technologies. We expect that foreign collection on US civilian and dual-use technologies will follow overall patterns of investment and trade. The following sectors—which are expected to experience surges in investment and are priorities for China—may be targeted more aggressively.

- *Clean technologies.* Energy-generating technologies that produce reduced carbon dioxide and other emissions will be the fastest growing investment sectors in nine of 11 countries recently surveyed by a US consulting company—a survey that included China, France, and India.
- *Advanced materials and manufacturing techniques.* One focus of China's 863 program is achieving mastery of key new materials and advanced manufacturing technologies to boost industrial competitiveness, particularly in the aviation and high-speed rail sectors. Russia and Iran have aggressive programs for developing and collecting on one specific area of advanced materials development: nanotechnology.

Rising Prices Increase Value of Commodity Information to Foreign Collectors (Index, 2002=100)

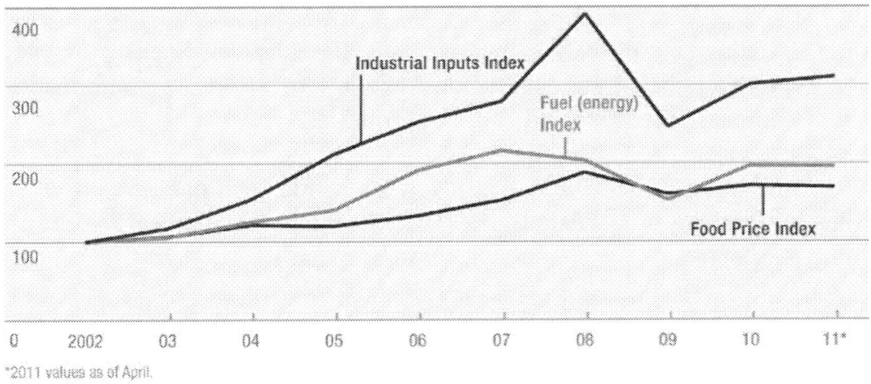

Source: International monetary Fund World Economic Outlook Database.

- *Healthcare, pharmaceuticals, and related technologies.* Healthcare services and medical devices/equipment will be two of the five fastest growing international investment sectors, according to a US consulting firm. The massive R&D costs for new products in these sectors—up to $1 billion for a single drug—the possibility of earning monopoly profits from a popular new pharmaceutical, and the growing need for medical care by aging populations in China, Russia, and elsewhere are likely to drive interest in collecting valuable US healthcare, pharmaceutical, and related information.

- *Agricultural technology.* Surging prices for food— which have increased by 70 percent since 2002, according to the food price index published by the International Monetary Fund (IMF)—and for other agricultural products may increase the value of and interest in collecting on US technologies related to crop production, such as genetic engineering, improved seeds, and fertilizer.[d]

Business information. As with technologies, we assess that nearly all categories of sensitive US economic information will be targeted by foreign entities, but the following sectors may be of greatest interest:

Energy and other natural resources. Surging prices for energy and industrial commodities—which have increased by 210 percent and 96 percent, respectively, since 2002 according to IMF indices— may make US company information on these resources priority targets for intelligence services and other collectors.[e]

- As noted earlier, cyber intrusions originating in China, but not necessarily attributed to the Chinese Government, since at least 2009 have targeted sensitive operational and project-financing information of US and other international oil, energy, and petrochemical companies, according to reports published by McAfee.

Business deals. Some foreign companies—at times helped by their home countries' intelligence services—will collect sensitive information from US economic actors that are negotiating contracts with or competing against them.

Macroeconomic information. In the wake of the global financial crisis of 2008-2009 and related volatility in the values of currencies and commodities, sensitive macroeconomic information held by the US private sector and government agencies is likely to remain a prime collection target for both intelligence services and foreign corporations. Chinese and Russian intelligence collectors may pursue, for example, non-public data on topics such as interest rate policy to support their policymakers' efforts to advance the role of their currencies and displace the dollar in international trade and finance. Such information also could help boost the performance of sovereign wealth funds controlled by governments like China's, whose China Investment Corporation managed more than $300 billion in investments as of late 2010.[f]

Possible Game Changers

Any of a range of less-likely developments over the next several years could increase the threat from economic espionage against US interests.

Emergence of new state threats. The relative threat to sensitive US economic information and technologies from different countries is likely to evolve as a function of international economic and political developments.

One or more fast-growing regional powers may judge that changes in its economic and political interests merit the risk of an aggressive program of espionage against US technologies and sensitive economic information.

Growing role of non-state and non-corporate actors. The migration of most business and technology development activities to cyberspace is making it easier for actors without the resources of a nation-state or a large corporation to become players in economic espionage. Such new actors may act as surrogates or contractors for intelligence services or major companies, or they could conduct espionage against sensitive US economic information and technology in pursuit of their own objectives.

Hackers for hire. Some intelligence services with less-developed cyber programs already use relationships with nominally independent hackers to augment their capabilities to target political and military information or to carry out operations against regime enemies. For example, the Iranian Cyber Army, a hacker group with links to the Iranian Government, has used social engineering techniques to obtain control over Internet domains and disrupt the political opposition, according to research conducted under an ONCIX contract. No evidence of involvement by independent hackers in economic espionage has been found in intelligence or academic reporting to date, in large part due to the absence of a profitable market for the resale of stolen information. This "cyber underground" could, however, become a fruitful recruiting ground for the tools and talents needed to support economic espionage. Following the model used by some intelligence services in exploiting the cyber environment for political or military espionage, a foreign government or corporation could build relationships with hackers for the development of customized malware or remote access exploits for the exfiltration of sensitive US economic or technology information.

Hacktivists. Political or social activists also may use the tools of economic espionage against US companies, agencies, or other entities. The self-styled whistleblowing group WikiLeaks has already published computer files provided by corporate insiders indicating allegedly illegal or unethical behavior at a Swiss bank, a Netherlands-based commodities company, and an international pharmaceutical trade association. LulzSec—another hacktivist group—has exfiltrated data from several businesses that it posted for public viewing on its website.

Corporate trade secrets or information about critical US technology may be at similar risk of disclosure to activist groups by disgruntled insiders.

- Antipoverty activists, for instance, could seek to publish the details of a new medicine under development by a US pharmaceutical company, with the goal of ending the firm's "monopoly" profits and making the product more widely available.
- Antiwar groups could disclose information about a new weapons system in the hope of dissuading the United States from deploying it.

ANNEX A. INTELLIGENCE COMMUNITY AND PRIVATE SECTOR MEASURES TO COUNTER ECONOMIC ESPIONAGE AND MANAGE COLLECTION IN CYBERSPACE

The IC is working closely with all segments of the public and private sectors to try to counter espionage activities that target our sensitive economic data and technology. We cannot expect to stop entirely or prevent hostile activity to collect US public and private sector information, but we can work to minimize the activity and mitigate its effects.

Intelligence Community Responses

The IC and especially counterintelligence (CI) officers have already taken a number of steps to improve collaboration, collection, and analysis across the CI, economics, and cyber disciplines.

Improved collaboration. Over the past few years, the IC has established multiple organizations and working groups to better understand the cyber espionage threat. These have contributed to a better understanding of the use of cyber in economic espionage.

- The National Cyber Counterintelligence Working Group established in 2011 is composed of 16 IC and other federal agencies and is creating a coordinated response to the cyber intelligence threat.

- The FBI is leading the National Cyber Investigative Joint Task Force, which brings together multiple agencies to collaborate on intrusions into US systems.

CI officers are considering an expansion of collaboration to include enhanced information sharing with Department of Justice attorneys. CI officers could introduce questions for attorneys to pose to offenders during the investigation process. They might also look at ways to tie plea bargains and sentencing decisions to suspects' willingness to cooperate with the CI Community during damage assessments.

Improved analysis and collection. The IC has made great strides over the past few years in understanding the cyber espionage threat to US Government systems, but our knowledge of cyberenabled economic espionage threats to the US private sector remains limited.

Defense Model Shows Limits to Mandatory Reporting Requirements

DoD's partnership with cleared defense contractors (CDCs) highlights difficulties in establishing an effective framework to improve the IC's understanding of foreign cyber threats and promote threat awareness in industry. The defense industrial base conducts $400 billion in business with the Pentagon each year and maintains a growing repository of government information and intellectual property on unclassified networks. CDCs are required to file reports of suspicious contacts indicative of foreign threats—including cyber—to their personnel, information, and technologies.

- Despite stringent reporting requirements for CDCs, DSS reports that only 10 percent of CDCs actually provide any sort of reporting in a given year.
- Another shortcoming of the defense model is that contractors do not always report theft of intellectual property unless it relates specifically to Pentagon contracts, according to outreach discussions with corporate officers.
- Corporate security officers also have noted that US Government reporting procedures are often cumbersome and redundant, with

military services and agencies such as DSS and the FBI often seeking the same information but in different formats.

Operations. CI professionals are adapting how they detect, deter, and disrupt collection activity in cyberspace because of the challenges in detecting the traditional indicators of collection activity— spotting, assessing, and recruiting.

It is imperative that we improve our ability to attribute technical and human activity in the cyber environment so that we can improve our understanding of the threat and our ability to generate a greater number of offensive CI responses.

Training and awareness. Expanding our national education and awareness campaign aimed at individuals and corporations is an essential defensive strategy for countering threats from cyber-enabled economic collection and espionage. We are building on current outreach initiatives that the FBI and ONCIX have already initiated.

- IC outreach to all US Government agencies, state and local governments, academia, nongovernmental organizations, industry associations, and companies is critical for promoting threat awareness, as well as for a better understanding of nongovernmental perspectives. Partners outside the IC are becoming aware of the wide range of potentially sensitive information in their possession and the extent of foreign efforts to acquire it.
- Outreach efforts include awareness and mitigation strategies for insider threat issues. The unique access of insiders to information technology systems and organizational processes makes this the most dangerous approach to cyber economic collection and espionage, as insiders can act alone to guide CNE or to download sensitive data to portable media.

ONCIX already engages in dialogue with ASIS International—an industry association for security professionals—and the Department of State's Overseas Security Advisory Council on the challenges facing both the public and private sectors with regard to cyber-enabled economic collection and espionage.

Finally, IC outreach efforts to the private sector on economic espionage need to fully engage corporate and other partners in order to be credible. We can facilitate partnerships to share best practices, threat updates and analysis, and data on intrusions.

One company security officer has suggested that the IC must speak to industry in language geared to the private sector's needs and experience and emphasize, for example, that the protection of trade secrets is critical to corporate profitability and growth.

As a follow-up to the public/private sector Workshop on Cyber-Enabled Economic Espionage held in 2010, ONCIX should consider sponsoring another conference with Department of Justice and private sector stakeholders on lessons learned regarding successful convictions under Section 1831 of the Economic Espionage Act.

Corporate Responses

The private sector already has a fiduciary duty to account for corporate risk and the bottom-line effects of data breaches, economic espionage, and loss or degradation of services.

A key responsibility of chief executive officers and boards of directors is to ensure that the protection of trade secrets and computer networks is an integral part of all corporate decisions and processes and that all managers—not just security and information systems officials—have a stake in the outcome.[a] Viewing network security and data protection as a business matter that has a significant impact on profitability will lead to more effective risk management and ensure that adequate resources are allocated to address cyber threats to companies.

- Only 5 percent of corporate chief financial officers are involved in network security matters, and just 13 percent of companies have a cross-functional cyber risk team that bridges the technical, financial, and other elements of a company, according to a 2010 study.

Judicial Mandate for Boards of Directors to Secure Corporate Information

Delaware's Court of Chancery ruled in the 1996 Caremark case that a director's good faith duty includes a duty to attempt to ensure that a corporate *information and reporting system exists and that failure to do so may render a director liable for losses caused by the illegal conduct of employees. The Delaware Supreme Court clarified this language in the 2006* Stone v. Ritter *case—deciding that directors may be liable for the damages resulting from legal violations committed by the employees of a corporation, if directors fail to implement a reporting system or controls or fail to monitor such systems.*

Companies that successfully manage the economic espionage threat realize and convey to their employees that threats to corporate data extend beyond company firewalls to include other locations where company data is moved or stored.

These include cloud sites, home computers, laptops, portable electronic devices, portable data assistants, and social networking sites.

- A survey of 200 information technology and security professionals in February 2011 revealed that 65 percent do not know what files and data leave their enterprise.
- According to a March 2011 press report, 57 percent of employees save work files to external devices on a weekly basis.
- E-mail systems are often less protected than databases yet contain vast quantities of stored data. E-mail remains one of the quickest and easiest ways for individuals to collaborate—and for intruders to enter a company's network and steal data.

Cyber threats to company information are compounded when employees access data through portable devices or network connections while traveling overseas.

Many FIS co-opt hotel staffs to allow access to portable devices left unattended in rooms. It is also much easier for FIS to monitor and exploit network connections within their own borders.

- Foreign collectors engage in virtual methods to collect sensitive corporate data and take advantage of victims' reluctance to report digital penetrations and low awareness of foreign targeting, according to legal academic research.

Corporate security officers have told ONCIX that US Government reporting procedures on economic espionage and cyber intrusions are often cumbersome and redundant.

Agencies such as DSS and the FBI often seek the same information but in different formats.

BEST PRACTICES IN DATA PROTECTION STRATEGIES AND DUE DILIGENCE FOR CORPORATIONS

Information Strategy

- Develop a "transparency strategy" that determines how closed or open the company needs to be based on the services provided.

Insider Threat Programs and Awareness

- Institute security training and awareness campaigns; convey threats to company information accessed through portable devices and when traveling abroad.
- Establish an insider threat program that consists of information technology-enabled threat detection, foreign travel and contact notifications, personnel security and evaluation, insider threat awareness and training, and reporting and analysis.
- Conduct background checks that vet users before providing them company information.
- Implement non-disclosure agreements with employees and business partners.
- Establish employee exit procedures; most employees who steal intellectual property commit the theft within one month of resignation.

Effective Data Management

- Get a handle on company data—not just in databases but also in e-mail messages, on individual computers, and as data objects in web portals; categorize and classify the data, and choose the most appropriate set of controls and markings for each class of data; identify which data should be kept and for how long. Understand that it is impossible to protect everything.
- Establish compartmentalized access programs to protect unique trade secrets and proprietary information; centralize intellectual property

data—which will make for better security and facilitate information sharing.

- Restrict distribution of sensitive data; establish a shared data infrastructure to reduce the quantity of data held by the organization and discourage unnecessary printing and reproduction.

Network Security, Auditing, and Monitoring

- Conduct real-time monitoring/auditing of the networks; maintain thorough records of who is accessing servers, and modifying, copying, deleting, or downloading files.
- Install software tools—content management, data loss prevention, network forensics—on individual computer workstations to protect files.
- Encrypt data on servers and password-protect company information.
- Incorporate multi-factor authentication measures—biometrics, PINs, and passwords combined with knowledge-based questions—to help verify users of information and computer systems.
- Create a formal corporate policy for mobility—develop measures for centrally controlling and monitoring which devices can be attached to corporate networks and systems and what data can be downloaded, uploaded, and stored on them.
- Formalize a social media policy for the company and implement strategies for minimizing data loss from on-line social networking.

Contingency Planning

- Establish a continuity of operations plan—back up data and systems; create disaster recovery plans; and plan for data breach contingencies.
- Conduct regular penetration testing of company infrastructure as well as of third-party shared service provider systems.
- Establish document creation, retention, and destruction policies.

Resources for Help

- Contact ONCIX or the FBI for assistance in developing effective data protection strategies.
  If a data breach is suspected, contact the FBI or other law enforcement/organizations for help in identifying and neutralizing the threat.

ANNEX B. WEST AND EAST ACCUSE CHINA AND RUSSIA OF ECONOMIC ESPIONAGE

Other advanced industrial countries principally blame China and Russia for economic espionage that results in large but uncertain monetary costs and job losses.

They perceive that China and Russia continue to use traditional human and technical collection methods—particularly against small- and medium-sized businesses—to gather economic information and technologies that save them research and development (R&D) resources and provide entrepreneurial and marketing advantage for their corporate sectors.

- Germany's Federal Office for the Protection of the Constitution (BfV) estimates that German companies lose \$28 billion-\$71 billion and 30,000-70,000 jobs per year from foreign economic espionage. Approximately 70 percent of all cases involve insiders.
- South Korea says that the costs from foreign economic espionage in 2008 were \$82 billion, up from \$26 billion in 2004. The South Koreans report that 60 percent of victims are small- and medium-sized businesses and that half of all economic espionage comes from China.[a]
- Japan's Ministry of Economy, Trade, and Industry conducted a survey of 625 manufacturing firms in late 2007 and found that more than 35 percent of those responding reported some form of technology loss. More than 60 percent of those leaks involved China.

France's Renault Affair Highlights Tendency to Blame China

Broad French concerns with Chinese economic espionage formed the background of the hasty—and subsequently retracted—accusations by corporate and political leaders in January 2011 that three top executives with the Renault automobile company had taken bribes from China in exchange for divulging technology.

- An investigation by the French internal security service revealed that the accusations against China lacked substance and may have stemmed from a corrupt corporate security officer's attempts to generate investigative work for a friend's consulting business.

Past Chinese economic espionage against the French automotive industry—including the parts manufacturer Valeo—probably made the French willing to give credence to any accusation of similar malfeasance against China.

Countries acknowledge the growing use of cyber tools for foreign economic collection and espionage and often note difficulties in understanding losses associated with these cyber collection methods. A 2010 survey of 200 industry executives from the power, oil, gas, and water sectors in 12 Western countries, China, and Russia indicates that 85 percent of respondents experienced network intrusions and that government-sponsored sabotage and espionage was the most often cited cyber threat.

- A 2010 Canadian Government report claimed that 86 percent of large Canadian corporations had been hit and that cyber espionage against the private sector had doubled in two years, according to a press report.
- The German BfV offers no reliable figures on the number of cases and amount of damage caused by cyber-enabled economic espionage, adding that their intelligence services are "groping in the dark." The German Government has noted the use of CNE tools and removable media devices, claiming that $99 million are spent annually for IT security.
- UK officials note that the cost of an information security incident averages between $16,000 and $32,000 for a small company and between $1.6 million and $3.2 million for firms with more than 500 employees. The United Kingdom estimates that attacks on computer

systems, including industrial espionage and theft of company trade secrets, cost the private sector $34 billion annually, of which more than 40 percent represents theft of intellectual property such as designs, formulas, and company secrets.

- Germany and South Korea judge that China, in particular, increasingly uses cyber tools to steal trade secrets and achieve plausible deniability, according to press reporting.[b]
- Unidentified CNE operators have accessed more than 150 computers at France's Finance Ministry since late 2010, exfiltrating and redirecting documents relating to the French G-20 presidency to Chinese sites, according to a press report.
- The British Security Service's Center for the Protection of National Infrastructure warned hundreds of UK business leaders in 2010 of Chinese economic espionage practices, including giving gifts of cameras and memory sticks equipped with cyber implants at trade fairs and exhibitions. This followed similar notification sent to 300 UK business leaders in 2007 warning them of a coordinated cyber espionage campaign against the British economy.
- German officials also noted that business travelers' laptops are often stolen during trips to China. The Germans in 2009 highlighted an insider case in which a Chinese citizen downloaded highly sensitive product data from the unidentified German company where he worked to 170 CDs.

China's Response to Allegations of Economic Espionage

China usually responds to public allegations of economic espionage with outright denial and counteraccusations. In 2009 China claimed the Australian mining giant Rio Tinto engaged in six years of espionage activities—bribery and information gathering—that resulted in a loss of iron ore imports for the Chinese steel industry as large as $107 billion. This loss was more than twice the total profits generated by the Chinese steel industry over that same six-year period, according to the Chinese Government.

Russia also is seen as an important actor in cyberenabled economic collection and espionage against other countries, albeit a distant second to China. Germany's BfV notes that Russia uses CNE and e-mail interception to save billions of dollars on R&D in the energy, information technology, telecommunications, aerospace, and security sectors.

- The Director-General of the British Security Service publicly stated that Russia, as well as China, is targeting the UK's financial system.
- A Russian automotive company bribed executives at South Korea's GM-Daewoo Auto and Technology to pass thousands of computer files on car engine and component designs in 2009, according to a press report.
- A German insider was convicted of economic espionage in 2008 for passing helicopter technology to the Russian SVR in exchange for $10,000. The insider communicated with his Russian handler through anonymous e-mail addresses.

Countries Suspect Each Other of Committing Economic Espionage

Allies often suspect each other of economic espionage—underlining how countries can be partners in traditional security matters yet competitors in business and trade. Foreign corporate leaders may make accusations that are not publicly endorsed by their governments.

- According to a 2010 press report, the Germans view France and the United States as the primary perpetrators of economic espionage "among friends."
- France's Central Directorate for Domestic Intelligence has called China and the United States the leading "hackers" of French businesses, according to a 2011 press report.

Some countries exercise various legislative, intelligence, and diplomatic options to respond to the threat of cyber-enabled economic collection and espionage.

- France and South Korea have proposed new legislation or changes to existing laws to help mitigate the effects of economic espionage. France also is considering a public economic intelligence policy and a classification system for business information.
- France, the United Kingdom, and Australia have issued strategies and revamped bureaucracies to better align resources against cyber and economic espionage threats. France created a 12-person Economic

Intelligence Office in 2009 to coordinate French corporate intelligence efforts. The United Kingdom established an Office of Cyber Security to coordinate Whitehall policy under a senior official and a Cyber Security Operations Centre within the Government Communications Headquarters (GCHQ) SIGINT unit. Australia created a cyber espionage branch within its Security Intelligence Organization in 2010.

- The United Kingdom is mobilizing its intelligence services to gather intelligence on potential threats and for operations against economic collection and espionage in cyberspace, according to press reports.

German Espionage Legislation Has Limited Results

Germany's Federal Prosecutor General initiated 31 preliminary proceedings on espionage in 2007, resulting in just one arrest and one conviction. German authorities note that espionage cases are often hindered by diplomatic immunity protections and by attribution issues from operating abroad through cyberspace.

Nearly all countries realize that public and private partnerships are crucial to managing the effects of cyber-enabled economic collection and espionage. The United Kingdom notes that 80 percent of its critical national infrastructure is owned and operated by the private sector. German authorities would like more corporate feedback and say that most enterprises either do not know when they are victims of cyber espionage or do not want to publicly admit their weaknesses. Most countries engage in some form of corporate outreach.

- The French intelligence services offer regular threat briefings to private companies, according to press reports.
- German authorities regularly exchange information with corporate security officers through a private/public working group that includes Daimler AG, Volkswagen, Porsche, Bayer, the German post office, and the railroad industry.

Corporate Leaders Speak Out on Chinese Espionage

Some foreign corporate executives have singled out Chinese espionage as a threat to their companies.

- British entrepreneur James Dyson—inventor of the bagless vacuum cleaner—warned in 2011 that Chinese students were stealing technological and scientific secrets from UK universities, according to a press report. He noted that Chinese students were also planting software bugs that would relay information to China even after their departure from the universities.
- The CEO of an Australian mining firm said that worries over Chinese and other corporate espionage drove him to adopt a more transparent quarterly pricing mechanism for commodities such as iron ore. He claimed that selling products at market-clearing prices visible to all would minimize the impact of differential information that one party may hold, according to a press article.

End Notes

[a] Chung was prosecuted only for possession of these documents with the intent to benefit the People's Republic of China (PRC) and acting as an unregistered foreign agent for China. He was not charged with communication of this information to the PRC or any other foreign entity.

[b] On average, one page of typed text holds 2 kilobytes (KB) of data; thus, 250,000 pages x 2 KB/page = 500,000 KB, or 488 megabytes (MB). A data CD with a capacity of 700 MB retails for $0.75, and a flashdrive with a capacity of 4 gigabytes costs about $13.00.

[c] An illegal is an officer or employee of an intelligence organization who is dispatched abroad and who has no overt connection with the intelligence organization with which he or she is connected or with the government operating that intelligence organization.

[d] The IMF's Food Price Index is a weighted index that includes the spot prices of cereal grains, vegetable oils and protein meals, meat, seafood, sugar, bananas, and oranges.

[e] The Fuel (energy) index published by the IMF is a weighted index that includes the spot prices of crude oil, natural gas, and coal. The Industrial Inputs Index is a weighted index that includes the spot price of agricultural raw materials (timber, fibers, rubber and hides) and non-precious metals (such as copper, aluminum, and iron ore).

[f] A sovereign wealth fund is a government investment fund, funded by foreign currency reserves but managed separately from official currency reserves. In other words, it is a pool of money that a government invests for profit.

End Note for Annex A

[a] Legal and human resources officers are two sets of key stakeholders given the role that corporate insiders have historically played in contributing to economic espionage and the theft of trade secrets.

End Note for Annex B

[a] We have no information on the methodologies that the Germans and South Koreans used to calculate their losses.

[b] We lack insight on the processes that the Germans and South Koreans used to attribute cyber activities to China.

INDEX

D

E

F

G

H

I

J

L

M

N

O

P

S

T